SCIENCE MYTHS
Unmasked

David Isaac Rudel

D1153803

Volume 1: Earth and Life Science

First published December 2010, ISBN: 978-1-935776-01-7

Interior artwork: Chanelle Henry and David Rudel
Cover design: Christy Bishop, Copen Marketing & Design
Substantive editor: Meagan Phillips
Copy editors: Rebecca Barns and Dr. Graham Stevens
Proofing: Mike Morrell, Jasmin Morrell and Dr. Graham Stevens
Fact checker: Dr. Graham Stevens

In addition to the above, the author thanks Dr. Charles Cunnigham, Dr. Paul Tjossem, Dr. Mark Schneider, Dr. Gerald Pollack, Dr. Peter Eastwell, and the editorial review board of the Science Education Review for advisory contributions.

The mask used on the cover is ©Gary S. Chapman.
The gadfly on the title page, from which Gadflower Press' logo is derived, is ©Alex Wild.
Feynman's quote (back cover) is from "Judging Books by Their Covers," a chapter in his famous *Surely You're Joking, Mr. Feynman!*

Contents

In fond memory of Stanley Leonard Miller.
He taught for 25 years at Stephenville High School,
enriching thousands of lives—including mine.

Introduction: Two Regrettable Forgeries

In 2005 I began working as an editor for ExploreLearning.com. We provide schools across North America with interactive educational simulations, called *Gizmos*™, designed to enhance standard science and math lessons. Because Gizmos are supplementary, we tailor them to align with the content regularly found in major textbooks. A surprising problem repeatedly arises during the creation of science Gizmos: the textbooks we review falsify the topic under consideration.

It can be said that textbooks are guilty of two regrettable *forgeries*, one matching each of the term's common meanings. In some cases textbooks forge counterfeit explanations, passing off myths as genuine accounts for natural phenomena. In other cases textbooks present factual information in misleading ways. If the human mind is a smithy wherein observation and experience shape one's beliefs about the natural world, then blights of this second type end up forging misconceptions when students (quite understandably) draw false conclusions from the tendered exposition.

Science textbooks perpetuate these myths and misconceptions because they are *efficient, effective,* and *expedient.* Claiming a candle burns out when placed under a jar because it has consumed all the available oxygen is effective for reminding students that oxygen plays a role in combustion. Claiming tides on Earth's far side (the half farthest from the Moon) are due to "centrifugal force" is an efficient method of explaining away a phenomenon whose legitimate treatment requires some challenging visualization. Claiming scientists approach problems using "the scientific method" is an expedient to help students develop certain cognitive skills pertaining to science.

Each of these three claims turns out to be doubly wrong. A candle consumes only a small portion of the oxygen in a jar, and it would go out even if oxygen were continually pumped into the vessel. Centrifugal force is purely illusory, so it should not be presented as the impetus behind any phenomenon—particularly not for tides since they would occur even if there were no circular motion in the Earth-Moon-Sun system. Lastly, "the scientific method" represents neither how scientists engage individual research problems, nor the long-term route by which science advances as a whole.

Emergency medical workers must triage incoming patients into categories: those who can be saved but require urgent care, those who can be saved without immediate aid, and those whose chance of survival is so small as to not warrant attention more productively spent on others. A similar demarcation has occurred in science education among those truths that are considered important and graspable, those that are elective, and those that will not appear on standardized testing. Textbooks often sacrifice honesty regarding members of this third group in their efforts to convey those deemed more important.

The problem is not that textbooks (or the education standards they address) treat a limited set of topics. American science education already suffers from "one inch deep and a mile wide" syndrome: I certainly do not advocate widening that trench. The problem is that publishers, in their zeal to hammer home lesson objectives, are willing to make untrue claims and present counterfeit accounts. Myths hamper comprehension of the key ideas they are intended to demystify and muddle students' conceptions of related, lower-priority topics. But they help pupils pick the right answer on standardized tests, so they persist.

Children are not the only victims of such information fraud. Teachers rely on textbooks to provide accurate accounts on a raft of topics. It would be absurd to expect teachers to personally verify their textbooks' presentations, nor are there convenient means to do so. As a former teacher, I can assure you the last thing frontline educators need is another demand on their time.

INTRODUCTION: TWO REGRETTABLE FORGERIES

For both of these victims—teachers who have been betrayed by textbooks and former students who were deceived by them—I've written these three volumes on common science myths and misconceptions. The work comprises[1] topics generally falling into one of four categories:

- Everyday observations explained by myths tied to a more important topic (e.g., describing the blood in veins as blue because it has no oxygen; misapplying Bernoulli's principle to explain flight)

- Fallacious portrayals of natural processes (e.g., claiming clouds form because cold air holds less water than hot; likening the warming effect of a planet's atmosphere to the operation of a greenhouse)

- Factual errors likely to cause larger misconceptions (e.g., misrepresenting the nature of science; providing the wrong definition for "producer")

---

[1] I am hopeful readers interested in this kind of book are also the type who will not mind a few remarks on *comprise* and *compose,* words I will frequently have need of in these volumes. Since the 1960s, "comprise" has increasingly been used as a synonym for "compose." It is becoming more and more common to read "oxygen and hydrogen comprise water" (or, put in passive voice, "water *is comprised of* oxygen and hydrogen"). The prevalence of this usage by careful writers is particularly odd, for it is forbidden by the *Chicago Manual of Style*, the *de facto* guide to American English. (Though I must admit that *Chicago* also frowns on long footnotes . . .)

The traditional meaning for "comprise" is "to include." (It comes from the same root as *comprehensive*.) Hence "comprise" and "compose" have historically been used as *antonyms*, their meanings converse to one another. That is to say "oxygen and hydrogen atoms compose a water molecule," and "a water molecule comprises oxygen and hydrogen atoms." This was the dominant use of "comprise" from the mid-18th century to the mid-20th century and its *only* use for over three centuries prior to that. Words are not static; their usage changes over time, but I do not know of any *verb* in English to have its meaning utterly reversed so that a pair of antonyms become synonyms.

- Factual errors so flagrant they are worth pointing out for their own sake (e.g., presenting simple machines as multiplying force; claiming that a substance stays at a constant temperature when it changes state)

In other words, I have tried to select serious issues rather than mere nitpicks. Most cause students to construct incorrect mental models of nature, cognitive frameworks that can hamper later studies and persist for a lifetime.

Beyond forging the aforementioned counterfeits and misconceptions, the emphasis on knowledge (rather than comprehension) encourages educators to *persuade* students of a claim rather than provide a sound reason for it. As an example, consider the Coriolis effect, which refers to the influence of Earth's rotation on the apparent motion of objects. In particular, an arrow moving straight (as seen from space) will seem to follow a curved path to observers anchored to Earth. Students are often asked to accept this phenomenon with reasoning similar to:

Other than the North and South Poles, all points on Earth rotate to the east. Points on the equator have the farthest to go in each 24-hour period, so they rotate the fastest. In general, the closer to the equator you are, the faster you rotate. This means than an arrow shot toward the equator flies over terrain rotating faster and faster eastward. To an observer standing on Earth's surface (rotating with it), such an arrow appears to curve because it travels over terrain moving eastward faster than it is.

This kind of reasoning may persuade someone that Earth's rotation causes flying objects to appear to curve, but the account is hardly robust. A student reading carefully has to wonder *what about an arrow shot due east?* Paths going eastward at all points are just lines of latitude on a globe. All points on such a path are rotating east at the same speed. According to the discussion given here, an arrow going due east should not curve at all, yet the Coriolis effect is equally potent in all directions. It is impossible to understand how the Coriolis effect can generate hurricanes if you believe it only influences the motion of objects moving north or south, which are the only objects textbooks cite when illustrating the topic.

Related to these efforts at persuasion are simple glosses of the form "Because A is true, B is true as well," where A and B are related, but not in the kind of logically tight way suggested by the sentence. For example, when explaining why you can rub a balloon against someone's hair and stick it to the wall, it is rather deceptive to say "because the balloon is negatively charged, it sticks to the wall." It is true that the balloon's negative charge leads to the attraction observed, but not in the straightforward manner described. The wall is neutral; it has no net charge. Negatively charged items generally have zero attraction to neutral ones.[2] A facile explanation like "because the balloon is negatively charged, it sticks to the wall" only encourages people to formulate erroneous beliefs about electrostatics. For example, a student could not be blamed for deducing that protons (positively charged particles) and neutrons (neutral particles) are held together in the nucleus of an atom for the same reason.

I believe I avoid falsely presenting such persuasions and glosses as genuine explanations in these chapters, though proper treatment of some topics demanded longer discussions than I would have liked. I have given suitable disclaimers in the rare instances that a satisfactory account was truly beyond the scope of secondary science education.

---

[2] In this case, the negatively charged balloon has the capability of causing the wall's distribution of charge to change, so that the surface of the wall is no longer neutral. See the *Circuits* chapter in volume 2 for more details.

My goal was to present accounts granular enough to allow complete understanding of the topic involved, and I provide notes, useful vocabulary, and sources for further study in an appendix. It is my hope that these volumes will find their way into the hands of many teachers. The detailed discussions should empower them to decide for themselves how to sculpt the material to match the needs of their students. Chapters on the simpler phenomena have been written in an informal tone in case educators wish to assign them as enrichment for students. The whole book should be readable by a precocious high schooler.

Should a reader find himself stumped by any of the descriptions given herein, I welcome requests for clarification as well as any other comments or suggestions. I am also looking for more examples of myths perpetuated by standard textbooks. Please send your personal favorites to `david@zukertort.com`.

1 The Seat Buckle Sting: Why do Metals Feel Hot?

After a summer day spent swimming, fishing, or people-watching, you return to your car and realize you forgot to crack the window. You open the door and hot, humid air rushes out. While you were gone, your car became a sauna.

As you sit down, the cloth seats are pleasantly warm against the backside of your knees—just one more perk of weather warm enough for shorts. You automatically grab the dangling seatbelt terminal near your shoulder and ZING! The seatbelt buckle gives your hand a much more painful welcome than the seat gave your legs. The seats reminded you of a mild heating pad; the metal seatbelt hasp brings to mind a frying pan.

Why does the metal burn you when the cloth does not?

When this question comes up in the classroom, often the explanation is: *Metals are conductors, so they heat up faster.* This narrative starts out on the right foot but then trundles into error—a different question should have been asked in the first place. Instead of asking why the buckle was hotter than the seat, we would do better to ponder why the buckle *felt* hotter than the seat. Your hand is not a thermometer. It cannot tell the temperature. Instead, nerves in your skin determine whether it is *gaining or losing heat.* (Scientists define "heat" differently than most people. *Heat*, strictly speaking, is thermal energy that is being transferred. It is not a reference to temperature. If this bothers you, simply pretend "heat energy" is written everywhere you see the word "heat.")

> Your hand is not a thermometer. It cannot tell the temperature. Nerves only determine whether the skin is gaining or losing heat.

Your internal body temperature is about 98.6° F (37° C), and your skin tends to stay around 91° F (33° C). *Because conduction moves heat from hotter objects to colder ones*, your body's metabolism is constantly warming your skin. It tends to lose energy to the outside air since you spend most of your time exposed to temperatures below 91° F (33° C). If the energy your skin is gaining from the inside is balanced by the amount it is losing to the outside, you are comfortable. However, if your skin cells are gaining thermal energy, your brain interprets this as "it's hot." Likewise, if your skin cells are losing thermal energy, your brain interprets this as "it's cold." This is how our nervous system reacts to short-term changes in our environment. Other factors, not discussed here, are involved for long-term, slow changes of temperature when core body temperature is affected.

> Heat flows from hotter objects to colder ones.

THE SEAT BUCKLE STING: WHY DO METALS FEEL HOT?

Ice Cube

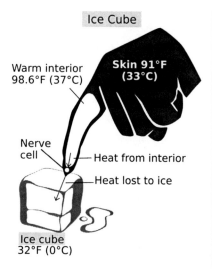

Warm interior
98.6°F (37°C)

Skin 91°F
(33°C)

Nerve
cell

Heat from interior

Heat lost to ice

Ice cube
32°F (0°C)

The ice pulls in more heat from
the surface of your finger than
is supplied by the warmth inside
your body, so it feels cold.

Room Temperature Air

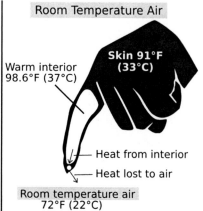

Skin 91°F
(33°C)

Warm interior
98.6°F (37°C)

Heat from interior

Heat lost to air

Room temperature air
72°F (22°C)

The heat lost by the skin to the
cooler air approximately equals
the heat supplied by the interior
of the finger, so there is no net
loss of heat. Even though the air
is cooler than your skin, the
nerve does not register it as cool.

Warm Air

Warm interior
98.6°F (37°C)

Heat from
interior

Skin 91°F
(33°C)

Warm air
91°F (33°C)

Coffee 150°F
(65°C)

The air is about the same
temperature as the skin, so no
heat is exchanged, but the skin
is still warmed by the inside, so
your brain thinks "warm," even
though no actual heat is entering
your skin from the outside air.

Hot Coffee

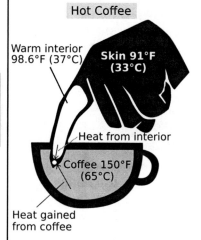

Warm interior
98.6°F (37°C)

Skin 91°F
(33°C)

Heat from interior

Coffee 150°F
(65°C)

Heat gained
from coffee

Your finger receives a great
deal of heat from the liquid
coffee and some from inside.
This could easily be enough to
cause damage to your skin, and
your brain registers "HOT!"

How Your Skin Feels Hot or Cold: Four Examples

This explains why it can be 86° F (30° C) outside and feel warm even though the air is cooler than your body! Your skin is losing a little energy to the air, which is slightly cooler. But your blood delivers more energy than is lost to the air. *Your skin is gaining more heat from the inside than it is losing to the outside, so you feel hot.*

The rate of heat loss (or gain) depends on both the temperature difference and the material you are touching. Your skin loses heat four times faster to water than to air at the same temperature.

The import or export of heat does not depend solely on the relative temperature, and that's where things get interesting. Heat flows from hotter objects to colder ones, but the compositions of the objects affect how quickly heat is transferred.

To understand how temperature and composition combine to determine the flow of energy, an analogy may help. Imagine a ball rolling down a hill. Temperature is represented by the ball's height in the analogy while the composition of the material is comparable to the roughness of the surface and the condition of the ball. Let's see how this works.

We know the ball tends to roll downhill rather than uphill. That is like knowing heat moves from hotter objects to colder ones, regardless of the materials in question. The temperature determines which direction the heat flows just like the slope determines which direction a ball rolls. Balls roll downhill. Period.

However, the ball's speed depends on both its composition and the steepness of the hill. A ball of putty will take longer to roll down the hill than a tennis ball. Balls roll more quickly down a steep hill than a gentle one, and they roll more slowly down a rough, grassy hill than a cement ramp. Thus, while the relative temperature determines which object gains thermal energy from the other, the *rate* of that flow also depends on the composition of the objects.

THE SEAT BUCKLE STING: WHY DO METALS FEEL HOT?

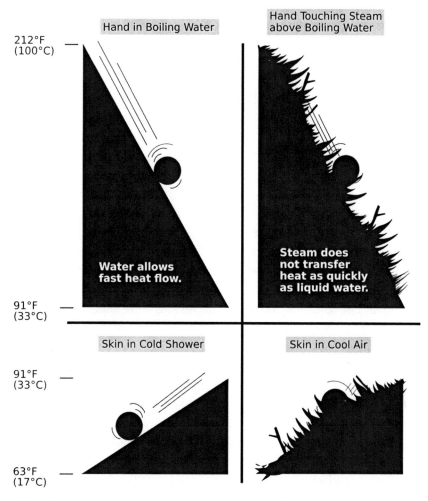

The relative temperature determines which direction heat flows, like the relative height determines which way a ball rolls down a mountain. But the **rate** of heat flow is determined by not just the difference in temperature but also the materials in contact. Some materials allow heat to flow very quickly, just as some surfaces allow balls to roll very quickly. Metals and water are analogous to concrete ramps in this regard while cloth and air are like grassy, bumpy slopes.

You can do a quick experiment to see this. Leave a glass of tap water sitting on your kitchen counter for half an hour. No matter what temperature the water was originally, it should be room temperature after being out that long. Put your finger in the water. The water feels cold, yet it is the same temperature as the glass, the counter, and the air around the rest of your body.

> Metal, like water, conveys heat to or from your skin very rapidly, so metals in a hot car burn you for the same reason that the first shower burst always feels cold.

To understand why the water feels colder than the air, recall that your skin loses heat to the air but gains heat from your internal metabolism. At comfortable temperatures these balance and your skin does not register a net change. However, your skin loses energy to water four times faster than to air, so your finger is losing much more energy to the water than the rest of your body is, more than the skin in your finger is gaining from the blood running through it.

The finger-in-water demonstration indicates why staying dry is so important when exposed to cold. Wet skin loses much more heat to a wintry environment. Also, it explains why the first burst of water from a shower head feels so frigid. The water in the pipes feels much colder than air of the same temperature.

How does this apply to the original question? Why does the seatbelt buckle burn you?

The "metals are conductors" part of the standard explanation is correct and relevant, but not because conductors heat up more quickly than insulators. Rather, metals feel hotter than cloth for the same reason that room temperature water feels much colder than room temperature air. Consider the metal in your oven at home. The metal trays are *heated by the air around them*, so their temperature cannot surpass the temperature of the air. Yet the trays feel much, much hotter than the surrounding air.

Even though the water and counter are the same temperature, your body gives up heat much more quickly to the liquid water than the wooden counter.

Conclusion

Metals typically conduct heat quickly, so touching them can cause rapid movement of energy, either into your body (if the metal is hotter than your skin) or from it (as when someone's tongue gets frozen to a flag pole). If the clasp on your car's safety belt is even a few degrees hotter than your skin, a great deal of heat can flow from the buckle to your hand. As to *why* the items in your car can get extremely hot, see the chapter on *The "Greenhouse Effect."* However, don't make the mistake of claiming the metal is actually hotter than the other items in your car. *Everything* will be hot in the car, but the metal will communicate that "hot-ness" much more quickly to your hand.

Addendum: Why Doesn't Foil Burn You?

As a side note to this topic, you might wonder why aluminum foil does not burn you. Indeed, I'm surprised this question does not come up more often as an oddity common to everyday experience. You can put a pizza in an oven set to 400° F (200° C) and ten minutes later tug on the foil without feeling any significant distress...unless your fingers graze the steel rack below, in which case your hand is likely to jerk back involuntarily.

Is aluminum an insulator instead of a conductor? No. In fact, aluminum generally conducts heat faster than the steel composing the oven rack. However, the foil is about 100 times thinner than the rack. This means the amount of heat available to send into your hand is extremely small. The foil does not hurt you for the same reason that a small ice fleck does not cool your cola much and a spark with a temperature of 8300° F (4600° C) poses little risk of burn.

Most explanations of the physics behind cloud formation either marginalize or omit altogether one critical fact: cloud formation depends on impurities in the air. Without that key piece of information, explanations must resort to erroneous claims and misapplied science.

This chapter describes problems with two common cloud-formation narratives. Thoroughly dissecting these stories makes the genuine cause of "ice cream castles in the air" pretty easy to understand.

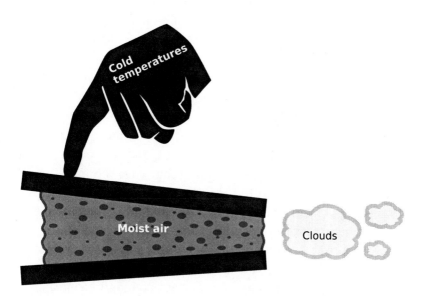

*How **not** to teach cloud formation—the idea that moist air loses water because it cannot hold it at lower temperatures was disproven in the 18th century.*

The Sugar-Dissolved-in-Tea Explanation

Students are sometimes asked to compare moisture in air to sugar dissolved in tea. More sugar can be dissolved into hot tea than cold tea. This fact is often phrased "hot tea can hold more sugar than cold tea." If you mix several spoonfuls of sugar into hot tea and let it cool, the sugar "un-dissolves" to form a clump on the bottom. Pupils are told water vapor "un-dissolves" from the air to form clouds for the same reason.

> Water vapor should never, ever be described as being dissolved in air. Nor should one say air "holds" water or is "saturated" with it.

Air naturally cools as it rises, and textbooks claim that water vapor condenses into minute water (or ice) particles when moist air no longer has the capacity to hold the water vapor dissolved in it. Sometimes textbooks even compare the air to a sponge.

John Dalton disproved this explanation over two hundred years ago. Its basic problem is simple: water vapor is not *dissolved*[1] in air at all! The oxygen, nitrogen, and other gases composing dry air—air minus the water vapor—have no material effect on cloud formation. One can reasonably speak of tea "holding" sugar because there is an actual interaction between the water molecules of the tea and the sugar molecules that prevents the sugar from gathering to form crystals. The molecules in hot water are more energetic than those in cold water, so they are better at keeping sugar molecules separated from one another. This allows hot tea to hold more sugar than cold tea. However, none of this has any connection to cloud formation.

To speak of air "holding" water would mean that air molecules interacted with the water particles, keeping them

---

[1] Sometimes "dissolve" is used in a much looser way than described here. My use is consistent with IUPAC—the world authority on terminology. Semantics aside, the important point is that air does not serve to separate water particles, so any analogy between hot tea and hot air is ill-conceived.

from gathering together into clouds. This does not happen. Nitrogen, oxygen, and other gases do not interact with water vapor in this way. Air does not act like tea, nor water like sugar. Indeed, any reference to "un-dissolving" when describing cloud formation is ludicrous.

> Relative humidity has nothing to do with cloud formation.

By definition, "dissolving" refers to mixing something into a *liquid* (or, occasionally, a solid), whereas air is composed of gases. You can speak of dissolving a gas into a liquid (e.g., the fizzy carbon dioxide in soda), but you cannot speak of dissolving *anything* into a gas. Molecules in a gas are too far apart to keep other molecules separated.

The "Overpopulation" Explanation

Some textbooks avoid all the nonsense about dissolving water vapor into air and instead simply declare that there is a maximum amount of water vapor that can exist in a given space at a particular temperature before condensation occurs, and this limit increases with temperature. The upshot of these claims is the same as the sugar-dissolved-in-tea explanation: water vapor condenses as it cools. Seldom is any reason given for why "overpopulation" causes condensation or why the upper limit changes with temperature. (This is probably why the "air holds water" narrative is so common: it provides a more persuasive case than "because we say so.") Students are asked to accept that the amount of water vapor that can exist in a volume

> If clouds formed simply from the overpopulation of water molecules, they would almost never be seen in our skies.

is limited and are told that *relative humidity* describes how close the actual amount in the air (the humidity) is to that bound.

11

This narrative views a sky region as a city whose denizens are water vapor molecules. The population density of the city is likened to relative humidity; when relative humidity reaches 100%, there is no room for the water vapor. Clouds, according to this description, comprise water droplets that cannot find housing, deportees from a crowded community. This explanation has a kernel of truth to it, but has to be rejected as unacceptable for three reasons:

1. It leads students to adopt an unrealistic picture of what is actually occurring in the sky. (One could say it not only messes up the "why" of clouds, but the "how" as well.)

2. It perpetuates a misunderstanding of relative humidity, a term having nothing at all to do with cloud formation.

3. While it is true that sheer "overpopulation" of water vapor could theoretically cause condensation, this is not the cause for cloud formation in our skies. Thus, the description fails in its basic goal.

The following three subsections address these points. Understanding the first of these problems is critical to grasping the second. Discussion of the second will lead to an appreciation of the fundamental role impurities play in cloud formation on Earth, which segues into the third.

1. The "how" of cloud formation

When I say the "overpopulation" explanation obscures the "how" of cloud formation, I'm referring to the incorrect implication that clouds form when water vapor is condensing and when there are no clouds it is because no condensation is occurring. Neither of these claims faithfully portrays reality. They are not merely wrong—they cause us to think in ways that hinder our ability to understand why clouds form in the first place.

Regardless of where you live or the outside temperature, *cloud drops* (minute specks of water or ice about 1/1000 the width of a raindrop) are constantly forming above your head. However, just because a cloud drop forms does not mean it will remain. These minuscule blobs of water can evaporate back into vapor as quickly as they form. In the air around us, water is always condensing

> Water evaporates and condenses microscopically in the sky above us all the time, even in deserts.

and evaporating microscopically. And so we see that cloud formation is not a question of whether condensation is occurring or not, but whether it is occurring *faster* than evaporation.

2. Relative humidity, misapplied

Once you understand that condensation and evaporation are constantly occurring everywhere in the sky, you can grasp the real problem with the "overpopulation" explanation. I will spend the rest of this section explaining the issue, but in short, *relative humidity has nothing to do with evaporation/condensation rates in **mid-air***. By definition, relative humidity is "relative" to the amount of vapor that can exist over a *flat surface of water*. Evaporation and condensation are occurring constantly, and at 100% relative humidity, the rate of evaporation from a lake's surface matches the rate of condensation onto it. However, none of this has any relevance to evaporation versus condensation in the sky, where the evaporation is not off a lake but from an inchoate cloud drop. Water evaporates much more readily from a tiny, curved droplet than from a flat surface because the forces holding the water molecules together are weaker.

In addition to the fact that molecules in curved drops are more prone to evaporation than those in a larger, flat surface, water vapor in the sky has difficulty forming drops at all without help. Colliding molecules tend to just bounce off one another due to their size and speed. Also, condensation heats up matter, for the same reason that evaporating sweat cools your skin. For condensation to occur, there has to be a place for that heat to

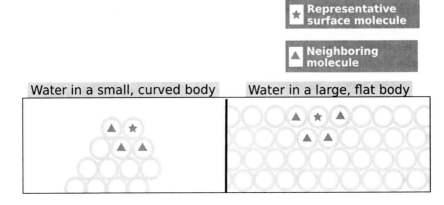

Water molecules on a small, curved surface have fewer neighboring molecules holding them in the drop than molecules in the surface of a large, flat body of water. The drawing under-represents this observation, for it only shows a two-dimensional version. In a three-dimensional array, a molecule on a flat surface can have nearly twice as many neighboring molecules than one on a tightly curved one. Water evaporates off small, curved droplets about three times faster than it does off large, flat bodies.

go. Water molecules have a hard time condensing unless there is some larger object to absorb the energy.

Because water vapor molecules have a hard time condensing from simple collisions, and it is so easy for water molecules to evaporate off a cloud droplet when it forms, an enormous amount of water vapor would have to crowd into a small volume before the particles collided with one another often enough for condensation to outstrip evaporation. For clouds to form in this way, due to sheer overcrowding of water vapor, the humidity in a region would have to get astronomically high, nearly four times what is currently needed to form clouds. We are lucky the white blotches in our skyscape are not caused by such overcrowding. Otherwise, few places on Earth would ever get rain, and those that did would be unbearably sweltering.

3. Pollution to the rescue

Luckily, we do not depend on spontaneous gatherings of crowded water vapor molecules to form clouds. Billions of particles float through the air, far too small to be seen by the naked eye. Water molecules have a much easier time sticking to these slower-moving, larger impurities than they do to each other.

Flecks of solid matter in the air are critical for cloud formation because water finds it much easier to come out of its gas phase if there is a solid substrate to adhere to.

Where do these particles come from? More or less everywhere! Almost any small scrap of solid matter can fit the bill: soot, dust, clay, sea salt, and even bacteria. They are extremely small (widths on the order of a millionth of an inch or 0.0001 mm). A cubic inch of air typically houses about 1,000 of these floating flecks. Each liter of air has about 100,000.

These particles give water molecules an anchor to cling to. Other water molecules can attach to those, until the layers form a cloud drop. These drops resist evaporation better than drops formed from simple collisions because they are larger. Bigger drops curve less than smaller drops. For this reason a nascent cloud drop is much more prone to evaporation than one that has reached its full size. According to C. F. Bohren's *Clouds in a Glass of Beer: Simple Experiments in Atmospheric Physics*, droplets have negligible curvature once they reach a width of 0.001 mm. Impurities in the air also lower the rate of evaporation by shielding water molecules that would otherwise be exposed on the boundary of the drop.

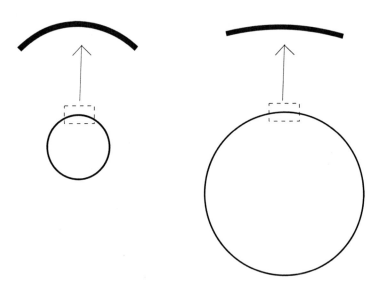

Smaller surfaces must curve more than larger ones. The figure shows magnified fragments from the edge of a small circle and a larger one. It might seem strange to say bigger circles curve less than smaller ones, but consider how standing on a beach ball compares with standing on Earth's surface. They are both (approximately) spheres, but Earth's surface seems flatter to someone standing on it.

Thus, pollutants actually help cause rain in two ways: (1) they give cloud drops an anchor for formation and (2) they lessen the likelihood that a cloud drop will evaporate once it has formed. Cities with more pollution tend to have less humidity because the specks of pollution make it easier for water vapor to condense.

If the impurities are cool enough, condensation will outstrip evaporation and cloud drops will form faster than they evaporate. Otherwise, the process of evaporation will restrain the number of drops, and no cloud forms. In neither case is the temperature of the *air* relevant, other than its effect on the temperature of the pollutants.

Conclusion

Clouds do not form because the air gets cold and cannot "hold" as much moisture, nor do they form because relative humidity reaches 100%. Clouds can form with or without air and at relative humidity levels above and below 100%. Relative humidity has no direct bearing on the amount of water vapor that can exist in a given volume prior to cloud formation.

Rather, clouds form because water vapor condenses on small flecks of matter in the air. A tug-of-war between the condensation and evaporation rates determines whether a cloud will form, and these rates are affected by a whole host of factors such as the density, type, and temperature of air pollutants and the amount of vapor in the region.

There is nothing technically wrong with the common presentation of Mendelian genetics[1] given in classrooms. The problem is that students are told, or at least given the strong impression, that inheritance generally follows the pattern exemplified by Mendel's peas. It doesn't. Our presentation of Mendelian genetics teaches students to think of genes and traits in ways that have little connection with reality. Moreover, the cognitive picture students construct based on Mendelian genetics is antagonistic toward the models they will study later regarding DNA and its purpose.

Why is there so much emphasis on teaching Mendelian genetics if it leads to such misconceptions? From a science education perspective, Mendelian genetics is a fantastic example of something interesting and not too hard to understand. Genetics is one of the few fields of middle-school science curriculum that has acreage in What-people-remember-after-graduation Land. There is a much better chance your blind date knows about recessive genes than Ohm's law.[2]

More specifically, Mendelian genetics allows teachers to easily show visible proof of something going on inside cells, or at least inside ours. (The vast majority of creatures, in particular most bacteria and fungi, don't have the cellular mechanics allowing for Mendelian genetics.) Mendel's discoveries give excellent support to the idea that we have two sets of genes, one inherited from each parent, and these work independently from one another rather than blending. A brown mouse mating with a white mouse does not necessarily lead to beige offspring. This is a landmark discovery in the history of biology.

---

[1]The term "Mendelian genetics" can be used in different ways. Here I refer to the version common to middle-school and high-school textbooks involving recessive and dominant genes, etc.

[2]Ohm's law relates the current in a circuit to voltage and resistance.

Furthermore, Mendel's work is an exemplar *par excellence* of scientific reasoning. Everything about his research program smacks of good scientific inquiry, except perhaps the part where he may have fudged his data.[3] Thus, there is certainly good reason to include Mendel's research in a discussion of genetics. The problem is that when Mendel's work dominates the classroom, as it does today, students are led to a false understanding of how most genes work (or don't).

Students come away with some variation on the following:

Each of our traits is determined by a gene (a stretch of DNA on one of our chromosomes). For each of these traits we have two genes, one inherited from our mother and the other from our father. There are two options (called alleles) for each gene, one is dominant; the other is recessive. If we inherit the dominant allele from either parent, it determines the trait. Otherwise, the recessive allele gets its way.

If the above were merely a stew of oversimplifications, it wouldn't be so bad. For example, there are typically more than two alleles for each gene. Textbooks focus on the traits where only two are prevalent to keep things simple. Such publishing decisions are easily defensible. A student who had been taught about wrinkled versus smooth peas and then found out later that some traits had three, four, or several options would not have to significantly modify her conception of genetics to assimilate the information.

---

[3] For years controversy has brewed over whether Mendel (or an assistant) cooked the books—the numbers in his experiments do not show the degree of variation we would expect in a fair experiment, at least at first blush. An investigation of this topic is extremely useful as it shows all the ways we humans can "find what we are looking for." *Ending the Mendel-Fisher Controversy* is a recent book with several essays by various voices on the matter.

But the problems run deeper. The common presentation of recessive and dominant alleles, the aspect of genetics that students best remember long after middle school, does the student a disservice. Few genes follow a dominant-versus-recessive scheme, yet it is portrayed as the typical, if not only, arrangement.

> Dominant and recessive alleles are relatively rare in nature, as are traits determined by a single gene.

A quick review of the linkage between genes and traits will prove useful. If you will permit an analogy, building a computer chip, car engine, or candy bar requires a *schematic*, a plan detailing the construction. For example, the schematics for a house are commonly called blueprints. Our most basic genes encode schematics our cells can use to synthesize proteins. These genes are called *structural genes* because they provide such construction guidelines. If a person's DNA does not include the schematic for a particular protein, the cells cannot build that protein. Since proteins are the basic building blocks for our tissues, it's natural that people whose genes encode different proteins will have different traits.

Once we understand how DNA affects our traits, it is not too hard to see why the dominant-versus-recessive model is false and, perhaps worse, can compromise a student's ability to grasp later material. Let's consider how each of these terms, "dominant" and "recessive," stack up when viewed within the gene → protein → trait regime.

> *Structural* genes encode schematics our cells can use to build proteins.

There is little reason *a priori* that the existence of one structural gene, the "dominant" one, should stop another gene from functioning. That would be like the schematic for a coffee pot jumping out of its file drawer, invading another, finding the plans for a tea kettle, and feeding

21

it into the office shredder. Drawings don't act that way, and neither do genes.

Cells use DNA as a resource, and the presence of one structural gene generally does not actively stop a cell's assembly line from utilizing another. In fact, there is a different set of genes, *regulatory genes*, that dictate which structural genes are used, but we normally don't discuss those genes in the context of recessive-versus-dominant genes.

> DNA is a resource for our cells, and the presence of one structural gene does not actively stop a cell from using another.

Similarly, "recessive" is an ill-chosen label for alleles. In general, the genes that we call recessive are not merely self-effacing—they are non-functional. It is not the case that a recessive gene actively causes one version of a trait while the dominant gene actively causes another. Most of the time, the recessive version does not do anything at all with regard to a given trait. If a mouse has a recessive gene we say causes white fur, it is far more likely that the gene in question does nothing whatsoever regarding the mouse's hair, and white fur is simply what you get when there is nothing to make it dark.

The view that recessive genes work unless there is a dominant one to stop them clashes with the model students learn in high school for the role of DNA. When a cell uses DNA to make proteins, the chromosomes are treated separately. The two do not compete for the cell's favor. A big improvement would be to cast heredity in terms of functional and non-functional genes, with the caveat that "non-functional" here means "not affecting the trait under discussion."

> Most "recessive" genes are not self-effacing—they are non-functional, at least with regard to the trait in question.

This would replace a model that actively causes students to develop false conceptions with one that merely indulges in a bit of oversimplification. Speaking in terms of functional and non-functional genes fits more smoothly into the high-school version of genetics. It also helps students more easily understand sex-linked traits—females are far less likely to get hemophilia because they have two chances to get a functional gene.

> It would be an improvement to stop speaking of alleles as "dominant" versus "recessive," and to instead speak of genes that are "functional" versus "non-functional" for the trait under discussion.

When we stop thinking of genes as matched pairs of dominant and recessive alleles, it makes it easier to understand why certain unrelated traits are genetically linked. A gene may be functional (it codes for a protein) while having no effect on the specific trait one happens to be discussing. A gene that we say "causes" white fur may actually produce a protein that does something completely different, like produce harder claws. That is to say, the section of DNA that potentially codes for dark fur overlaps with the section that potentially codes for the harder claws. Each chromosome could have only one gene or the other, so a mouse that gets two of the hard claw genes would necessarily get none of the dark fur genes.

A useful example to consider is the gene that determines whether someone has sickle cell anemia. One allele codes for hemoglobin that can cause sickle cells but is resistant to malaria while the other codes for hemoglobin more vulnerable to malaria but not prone to sickle cell. In this case, neither gene can be considered recessive, they just code for different things. Someone who inherits one copy of each gene makes some hemoglobin of each kind. He will produce a few sickle cells, but not enough to cause concern. The gene that causes sickle cell anemia only looks recessive when we force the topic into our Mendelian model.

I have spent most of this chapter describing regrettable consequences of the dominant-versus-recessive model. A more basic pitfall lies in believing each trait is determined by a single gene. The connection between our genes and our traits is neither as direct nor as pervasive as students are led to believe. It is extremely rare for a trait to be controlled by a single gene. There are a few genes, called *invariant genes*, that operate that way, but they are very much the exception and generally relate to dire problems, like sickle cell. Most traits are either influenced by an ensemble of genes (each of which can play a hand in a variety of traits) or not significantly affected by genetic disposition at all.

By itself, this oversimplification would be just that: an over-simplification. Its effect is compounded through the natural connection between genetics and evolution. *Naïveté* in one can lead to something worse in the other.

Conclusion

Textbooks lead students to believe that most traits are ac-tively caused by particular genes. For example, white fur is effected by one gene while brown fur is brought about by an-other. In this regime, one of the rival alleles "dominates" the other in those instances when a mouse inherits a copy of each.

This dominant-versus-recessive model is not applicable to most traits. Perhaps of greater concern, these abstractions cause students to develop conceptions antagonistic to the mod-els they will learn in later biology classes. Furthermore, as we will see in the next chapter, the deficiencies in our treatment of genetics drastically limit our ability to present evolution in an honest fashion. Textbooks paint themselves into a corner by pressing for an understanding of evolution not commensurate in sophistication with their treatment of genetics.

Middle-school texts almost always do readers a tremendous disservice as they move from genetics to evolution. Of course, there is no inherent error in linking genetics to evolution. One is hard pressed to find two biology topics more naturally related. At a practical level, though, good pedagogy tends to crash headlong into a paradox that paralyzed the scientific community a hundred years ago. Mendelian genetics suggests biological stability and appears to forbid gradual[1] changes; traditional evolutionary theory (called neo-Darwinism) contends the opposite.

Pre-Neo-Darwinism

To grasp the significance of the problem in moving from Mendelian genetics to evolution, a bit of history is useful.

Around the turn of the 20th century, scientists hoping for a gene-based theory of evolution were flummoxed by the very Mendelian genetics we now teach students as supportive of evolution. If creatures inherit genes from their parents, and those genes are immutable, how could a species acquire new traits? For example, imagine that bird beaks were controlled by a gene that either caused long beaks or short beaks in a population. From a Mendelian perspective, beaks could not evolve to longer and longer lengths because the best genetic makeup any specimen could have is two copies of the long-beak gene. Other than random fluctuations that had nothing to do with genetics, there would be no way that a population's average beak length could break through this genetic glass ceiling.

---

[1]Note that throughout this chapter "gradual" primarily refers not to the *rate* of overall change but rather that the changes can occur in extremely small steps. This requires a smooth *spectrum* of possible options.

To get around this problem, scientists hypothesized that *mutations* are also heritable. This requires the immutable genes of Mendelian genetics to be not so immutable after all. In 1908, Thomas H. Morgan[2] showed that mutations could arise and be inherited.

During the early 20th century, Mendelian genetics presented an obstacle to Darwinism. Mendel's work indicated change could only occur abruptly, but Darwinism called for gradual transformation.

Surprisingly, this led to a much greater problem for those who wanted to follow Darwin's version of evolution by natural selection. Mutations allowed for new traits to enter a population, but not in the gradual way Darwin suggested. The new gene would not *blend* with the others, so only abrupt changes were allowed. Hugo de Vries suggested such a theory of evolution. His theory was compatible with known genetics, but not supported by fossil evidence (which showed gradual changes in at least a few places) or natural observations (where gradual variation was common). Morgan's research was meant to find support for de Vries' theory, and his results did just that.

Morgan pointed out another problem with addressing the deficiencies of Darwinian evolution by appealing to mutations. In his Princeton lectures, published as *A Critique of the Theory of Evolution* in 1916, he indicated that the *immutability* and *persistence* of genetic factors allowed only for creative evolution—instances of new forms. While "the survival of the fittest" might explain why injurious mutations never catch on, it could not explain how a species *as a whole* could change. An

---

[2]Morgan became a superstar in the field of genetics. In addition to showing that mutations could be inherited, he also fathered the chromosome-based theory of genetics, co-authoring the seminal textbook on the subject in 1915. C.H. Waddington compared his work to Galileo's or Newton's. He was in a perfect position to evaluate the evolutionary theories of his day.

advantageous mutation might appear in a population and persist as a new type of organism, but the claim that the mutant version could dominate the population, choking out the original viable creature, was not compelling.

In this view, evolution could beget new species but not transform older ones. In other words, Morgan disagreed with "survival of (only) the fittest," which was needed to support the lineages presented by Darwinists showing a progression where each species was superseded by the next.

I describe more of this theory's history in the final *Nature of Science* chapter (see volume 3), but suffice it to say that these objections hobbled the field for decades. The problem was not merely with Darwin's theory *per se.* Rather, the evidence used to support evolution suggested species change gradually over time. However, Mendelian genetics appeared to forbid this unhurried gait. There appeared only three options, none very satisfying:

1. Accept gradual evolution, but assume it was not based on genetics or heredity. Proponents of this view had a difficult time explaining how evolution could occur at all.

2. Claim evolution was based on genetics and therefore occurred in large, distinct steps, as Mendel's work showed. Proponents of this view had difficulty explaining why nature showed several instances of gradual variation.

3. Claim evolution was based on genetics but that it also occurred gradually, regardless of Mendel's work. Proponents of this view had to rationalize their acceptance of two theories that conflicted with each other.

In short, Mendel's work made it difficult to understand how evolution could be *genetic.* It would be decades before fancy mathematics showed that a new formulation—where a collection of genes inform most traits (*polygenetic inheritance*)—allowed for a genetic, gradual theory of evolution.

The Plight of the Educator

The historical problems outlined above give some indication of the inherent difficulty faced by middle-school teachers and the writers of their textbooks. Trying to go from a chapter on Mendelian genetics to a chapter explaining natural selection must *ipso facto* gloss over obvious concerns it took professional naturalists four decades to circumnavigate, and then only because mathematics rescued them from a sinking ship.

Textbooks naturally use transitions that try to show how our understanding of heredity substantiates the neo-Darwinian theory of evolution. The problem is that the *version* of Mendelian genetics we teach definitely does not support the version of evolution we teach, especially in middle school. This makes a satisfactory transition from heredity to evolution impossible.

Several specific indictments against textbooks can be made. I summarize them here before expanding on each:

1. They fail to discuss the critical importance of polygenetic inheritance with multiple alleles.

2. They do not address the problems Mendelian genetics poses for the evolution of invariant genes, genes that directly and single-handedly cause particular traits.

3. They use misleading examples in an effort to gloss over the impossible transition they ask students to make.

4. Because they give so little emphasis to the role of mutation, they end up avoiding altogether the problem of species transformation cited by Morgan.

1. Polygenetic inheritance and multiple alleles

Most inheritable characteristics are influenced by several genes acting in concert, a phenomenon labeled *polygenetic inheritance*, yet many middle-school textbooks do not discuss it. Furthermore, those that treat it with a paragraph or two fail to describe the critical role it plays in harmonizing a theory of gradual evolution with the abruptness of Mendelian genetics.

A student reading a typical chapter on evolution in a middle-school book has every reason to wonder how it is possible that genetics could allow the type of infinitesimal changes described. In Mendel's experiments, peas were either smooth or wrinkled, they were never semi-smooth. Polygenetic inheritance allows for greater variability but is not the only alteration of Mendelian genetics required to provide the gradations we see in nature. We also have to get away from the two-allele, dominant-versus-recessive model. To see why that model fails, let's try to apply it to human skin color.

Some statistical analysis suggests human skin color is determined by three or four genes acting in concert. For ease of computation, we will assume three genes are responsible. It does not matter which genes are dominant, so let's say darker skin is dominant over lighter skin. This means the palest people have no dominant versions for any of the genes. We could denote this category $\langle R, R, R \rangle$, meaning the person had only recessive versions of each gene. The darkest people would have at least one dominant allele for each gene, which we would denote $\langle D, D, D \rangle$. Someone who had a dominant allele for the first and second gene but only recessive alleles for the third would be $\langle D, D, R \rangle$ in this notation. Since the genes may have different influence, we must keep track of them separately.

Each of these categories, which scientists call *phenotypes*, refers to an observable genetic possibility. It is not hard to see that there are eight possible categories. Obviously, there are far more than eight categories for skin color.

Let's look at what happens when we discard the notions of dominant and recessive. Now there are *three* options for each gene: You could have two copies of one of the alleles, you could have two copies of the other, or you could have one copy of each. Back when we thought in terms of dominant and recessive, having two copies of the dominant allele was the same as having one of each, but now those are considered different. Instead of $2 \times 2 \times 2 = 8$ divisions, we now have $3 \times 3 \times 3 = 27$. This is better, but still not high enough to explain all the differences we see.

However, if we then allow for *three* alleles for each gene, we arrive at six options for each gene. There are three ways to get a pair of the same allele, and three other ways to get two different alleles. This means we have 6 x 6 x 6 = 216 different genetic classes! Obviously, this allows for more gradual changes between individual genetic categories than the model for genetics most people are taught.

The figure on the facing page compares the first three of these scenarios. For the last, a model called *incomplete dominance* is used. It is the simplest, but by no means the only, way for nature to deviate from what is shown as standard in textbooks.

Since the world's brightest biologists were stumped by this problem for four decades, omitting it from a discussion of evolution through genetics glosses over an important point. We should be asking students to be critical readers and thinkers, but we have them read textbooks that are written with the expectation that they won't question or consider deeply any of the explanations given.

2. Invariant genes

From a pedagogical standpoint, if not from a scientific one, the issue of invariant genes is even more obvious. Since some traits are clearly controlled by a single gene, how can that single trait change gradually from one expression to another? Students are taught to think of this kind of heredity as the default. "How

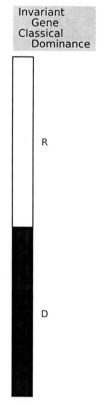

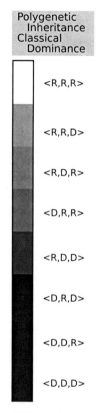

Invariant
Gene
Classical
Dominance

R

D

Polygenetic
Inheritance
Classical
Dominance

<R,R,R>

<R,R,D>

<R,D,R>

<D,R,R>

<R,D,D>

<D,R,D>

<D,D,R>

<D,D,D>

Polygenetic
Inheritance
Incomplete
Dominance

<0,0,0>
<0,0,1>
<0,1,0>
<0,0,2>
<0,2,0>
<1,0,0>
<0,1,1>
<2,0,0>
<0,1,2>
<0,2,1>
<1,0,1>
<1,1,0>
<0,2,2>
<1,0,2>
<2,0,1>
<1,2,0>
<2,1,0>
<2,0,2>
<2,2,0>
<1,1,1>
<1,1,2>
<1,2,1>
<1,2,2>
<2,1,1>
<2,1,2>
<2,2,1>
<2,2,2>

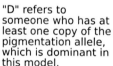

"D" refers to someone who has at least one copy of the pigmentation allele, which is dominant in this model.

This model assumes three separate genes inform skin pigmentation. In each case, the allele for pigmentation is considered dominant.

<D,R,D> refers to someone who has the allele for pigmentation at the first and third genes, but two copies of the recessive allele at the second gene.

This model assumes that having a single pigmentation allele at a given gene causes lighter skin than when both alleles at the gene are for pigmentation.

<2,0,1> refers to someone with two pigmentation alleles at the first gene, none at the second, and one at the third.

does a species change gradually when there are only two (or a few) options for each trait?" should be an obvious question, yet our textbooks tend not to address it at all.

3. Misleading examples

Textbook publishers must use special ink when printing chapters on evolution, an ink including peculiar nectar that attracts flawed examples. I do not know of a field that regularly is illustrated with more ill-conceived exemplars than natural selection's role in evolution.

One such standard example is Kettlewell's study of peppered moths in the mid-20th century. He documented a change in the prevailing color of the peppered moth from light to dark around the time of the industrial revolution and a return to a paler hue around the time regulations cut down on pollution. Kettlewell attempted to show that this change was due to the better camouflage afforded by the darker version during those decades when heavy pollution stained trees black with soot. A host of controversy has erupted over the methods, techniques, and conclusions of this study, not the least of which being that the geographic distribution of darker moths did not match the geographic distribution of discolored trees.[3] But the main reason it doesn't belong in a middle-school book on *evolution* is that it portrays a fatally crippled, untenable version of neo-Darwinism.

The type of evolution seen in this study could never cause any new forms or even a change in species. Kettlewell's moths show *natural selection* at work, but they are included in chapters devoted to long-term evolution to explain how new species arise from old. To insinuate that Kettlewell's moths show how

---

[3] Many papers and a couple of books have been written on these experiments, and battle lines have been drawn in the biological community. I won't comment further here on the specifics, as the charges are legion and severe. However, it seems reliable that, whatever errors or fraud existed in the original study, there is certainly *some type* of natural selection involved. However, it is not at all clear that the reason for the selection presented by textbooks (effect of bird predation on camouflage) is correct.

such long-term evolution comes about is fundamentally dishonest. This is not a matter of elapsed time; the processes seen in Kettlewell's study could never lead to a new moth form, even if given a billion years. It is not a case of "microevolution" versus "macroevolution," a fuzzy distinction at best, but of "evolution without mutation" versus "evolution with mutation." The fundamental claim of neo-Darwinism is that *genetic mutations* account for the evolution of today's complex lifeforms from the simplest cells, assuming certain rules for population genetics. Kettlewell's moths only distract from this key thesis.

Another common illustration in textbooks presents a collection of tooth fossils showing a gradual change in form. One obvious (unanswered) question is how this change could occur within the genetics framework students have just learned. When Mendel bred wrinkled peas with wrinkled peas, he did not get a batch that was *extra* wrinkled. Mendelian genetics suggests there is some upper bound on how extreme a given trait can become. Individuals within a population might become more and more likely to have that extreme case due to selection, but simply having more specimens with the most extreme genetic makeup possible would not allow any to go beyond that limit.

The above objection is, of course, just another manifestation of the glass-ceiling effect mentioned earlier (the beak discussion on page 25), an objection Morgan himself raised. However, in the case of teeth there is a much more significant problem. The teeth a creature grows are greatly affected by non-genetic factors. I discuss this more later, but the type of tooth an animal develops is largely determined by the kind of seeds it eats in its formative years. Thus, animals may indeed change their forms over time (if the environment changes from one generation to another) without this change saying anything about the underlying genetics. A vast array of tooth fossils can be presented purportedly showing a species evolving over time when in reality no *genetic* changes were occurring at all.

One final class of bad examples bears mention. If a textbook is trying to find legitimate mutations that cause real change in a species over time, they might cite cases of bacteria that mutate so they can eat new substances. For example, bacteria have been cultured so that a mutation allows them to digest lactose.

If you are looking for real laboratory work showing genuine evolution through mutation and natural selection, these are probably the best ones you'll find in modern textbooks. However, even they are fundamentally flawed. The kind of mutations in these examples cannot be responsible for long-term evolution.

Bacteria that evolve to eat lactose[4] do so by a change in their *regulatory genes*, not their structural genes. Regulatory genes determine which genes cells use. When a bacterium population mutates to digest lactose, it is not because it has added a new schematic to its filing drawer, a new protein it can make that earlier it couldn't. Instead, the bacteria have either dusted off blueprints they already had (but were not using) or retired some they were using that are now no longer useful. The bacteria did not gain a new ability but activated a capability it already had.

Such mutations alone are unable to evolve more complicated life forms from simpler ones because they do not increase a cell's arsenal of schematics. A mutation that merely allows or forbids an organism to use the DNA it already has is qualitatively different from a mutation that changes the cell's library of protein blueprints. It is the difference between flicking a light switch and inventing the light bulb.

It would be nice if textbooks used examples of evolution where a mutation brought about a new *structural gene* that improved the fitness of the organism. This seems like a simple enough request, as it is only asking for an example exhibiting the type of small changes that neo-Darwinism depends on.

---

[4] Bacteria have also been cultured to break down nylon, but this is also not suitable as an example of traditional neo-Darwinism. See page 45.

4. Survival of (only) the fittest

The final issue involves the subtle difference between one species *giving rise to* another versus a species *transforming into* another. After seventy years of experts' debating the topic, Morgan was completely unconvinced on the matter. Contrary to Darwin's view, which Spencer articulated as "survival of the fittest," Morgan declared, "evolution assumes a more peaceful aspect" rather than the "ferocious struggle between the individuals of a species with the survival of the fittest and the annihilation of the less fit." This led him, in the conclusion of his series of lectures on the topic, to claim *natural selection* could only refer to the increase in the number of individuals having a certain beneficial trait (rather than the obliteration of those that lack this trait) and the increased probability of future changes expected once this new population had a foothold.

Morgan may have arrived at this conclusion due to several independent lines of evidence, but the argument developed in his Princeton lectures involved how mutations accumulate and spread in populations. Luckily, there are easier ways of seeing how natural selection is limited in its ability to achieve such *destructive* evolution, the obliteration of one species as it is transformed to another.

You don't have to be a Princeton-caliber student attending lectures by an eminent evolutionary biologist to see the problems here, but the version of genetics taught to students makes even graspable issues opaque. In a world where genes are either dominant or recessive, it is not hard to see how a recessive, helpful gene could gradually overtake a population *in toto*. All specimens that have the gene for the dominant trait are at a disadvantage. In other words, the less fit version has no place to hide because it is dominant. It is a little harder to see how a dominant gene could completely overtake a population, allowing for a complete transformation of species. The recessive genes can hide behind the dominant ones, persisting in the population even when very few show the unfit phenotype.[5] Even in this case, however, students can easily come to the conclusion that the beneficial genotype will at least gradually increase in proportion. Having considered those two cases, students can freely conclude populations move inexorably to the best genotype available.

Natural selection is not always mathematically permitted to favor the optimal genotype.

However, in a world where most genes do not fit the dominant-recessive model, there is the very real issue that *the most-fit individual may not be genetically selectable!* Consider the sickle cell gene I mentioned earlier. The "fittest" individuals will be *heterozygous* for this gene, meaning they have one of each allele. These people are unlikely to get sickle cell anemia, and they are also resistant to malaria. Since the most advantageous genotype is to have one allele of each type, the "survival of the fittest" can never favor one allele over the other.

Not only is natural selection unable to favor one allele over the other, it is also mathematically handicapped when it comes to favoring the optimal combination. This is a simple application of probability. As illustrated on the facing page, no matter whom a heterozygous individual mates with, no more than half of her offspring are expected to match her genotype. Unless you are in the extreme case where the heterozygous genotype is not merely favored but is the *only* viable genotype (e.g., all others die as infants or are sterile), it is mathematically impossible for natural selection to transform the population to one where the favored genotype is standard.

---

[5]This reasoning only applies to the question of one gene overtaking a population, allowing the species to transform by completely replacing the other. It turns out that the *opposite* logic is true when determining whether a mutation manages to get a foothold in the population at all. Dominant genes have an easier time doing this than recessive ones.

Father is AB			Father is AC			Father is BB			Father is CC		

Below is the set of Punnett squares. Mother is AB on the left axis (rows A and B).

Father is AB (columns A | B):

	A	B
A	AA	AB
B	AB	BB

Father is AC (columns A | C):

	A	C
A	AA	AC
B	AB	BC

Father is BB (columns B | B):

	B	B
A	AB	AB
B	BB	BB

Father is CC (columns C | C):

	C	C
A	AC	AC
B	BC	BC

At most half of all offspring of a heterozygous parent will match the parent's genotype. The above illustration captures 4 of the possible 6 cases if one assumes 3 alleles for a given trait.

This common situation, where the optimal genetic makeup involves having two different alleles, is known as *heterozygous advantage*, and it is just one example of when natural selection cannot select for the best genotype. Oddly, the reverse situation has an even stronger effect. Imagine a mutant gene arises in a population that is mildly harmful if inherited from just one parent but extremely beneficial if inherited from both. A little thought experiment will show that the most fit genotype (the one with two copies of the mutant gene) will not only stay limited, but will never take hold at all.

Outdated Information

There are other unsettling choices made in evolution education. Many would have been defensible 50 years ago, but now betray general neglect on the part of textbook writers. The original version of neo-Darwinism came into existence before many modern subfields of biology even existed. Furthermore, for reasons that are still not well understood by those who study the history of science, the commonly articulated version of evolution today is actually *less* broad than its original form.

The original version of neo-Darwinism was a triumphant application of mathematics to genetics that allowed for a *gene-*

based theory of evolution. Natural selection was only one of several mechanisms that were all assumed to work together within a certain mathematical framework. Complex mathematics using this framework showed that Mendelian genetics did not have to be opposed to evolution. However, the next fifteen years brought a surprising shift in the *articulation* of the theory. Biologists began ignoring all factors other than natural selection. The other mechanisms, critical components in the mathematical logic originally justifying the theory, were mostly thrown out once the mathematicians had left the discussion.

More information on this surprising shift can be found in the final chapter of volume 3. Here I'd like instead to point out the increasingly clear difficulty in adhering to this version of evolution. The issue is not whether natural selection plays a role in evolution, nor is it that genes are important to evolution. The question is "How central are these aspects, as understood in neo-Darwinism, to the story of evolution on Earth?"

Asking a scientifically literate person this question could well lead to a blank face. It is not at all obvious, given the desire for change to be hereditary, how there could be any evolution that was *not* gene-based. Furthermore, it is hard to understand how random forces could have anywhere near the impact natural selection does in shaping evolutionary history.

With regard to the first concern, we are not weighing gene-based evolution versus non-gene-based evolution. The concern is rather that genes may have roles quite afield from the one they play in neo-Darwinism, and it isn't clear which of their various roles have the greatest influence on evolution. I present specific examples in the next section.

I also present examples showing how mechanisms other than natural selection shape evolution, but first a clarification: saying that evolution is due to something other than natural selection *is not* the same as saying it is "random," at least in the way that word might usually be understood. The idea behind natural selection is that one sub-population *has an advantage over another* and so this sub-population should increase in size while the other decreases. However, this line of reasoning assumes that a particular change produces a new version that *strives with the main population*. The "survival of the fittest," by which is really meant "survival of (only) the fittest,"

> Natural selection only works when there is competition. This is one reason it requires gradual changes, because the mutant form of a species must be close enough to the original to compete with it.

only works when species have to fight for limited resources or are otherwise occupying overlapping niches. If a genetic change (whether *beneficial* or not) renders an organism different enough that it fills an independent niche, one can no longer talk intelligently about "relative fitness" or nature "selecting" one creature over another.

A hypothetical scenario illustrates this important point well. Assume a population of birds has variable beak length, and they are able to eat seeds A, B, and C on a given island. They are unable to eat seed D for some morphological reason. Perhaps it is too big to swallow and too hard to crack. Suppose a mutation arises that allows its bearer to eat seed D but the mutation makes it impossible to eat seeds A, B, and C. If seed D is less bountiful than the other seeds, it is hard to say this mutation increases the fitness of the bird. However, if there are no other organisms eating seed D, the likelihood that the particular mutant will survive is large. One can easily see various ways this can play out depending on how the mutant's offspring work with the original set, but the point is that in no abstract sense can this phenotype be considered "more fit" than others. The mutant in question has no problem competing (if that is the correct term) simply because it has no competition for a time. The population it gives rise to may be small, but it should survive all the same.

The above is a pure case. It's more likely that the mutant *partially* competes with the original population. The effect is the same. Since the competition is limited, a new species may come to exist even though it is in no sense more fit than the original. It simply occupies a different niche.

Of course, one could make the case that such large changes are rare, but another could claim in response that their magnitude more than makes up for their infrequency. A single such mutation could cause greater evolutionary change than thousands of generations of the population acting only under the much more gradual changes amenable to natural selection.

Examples of Non-Standard Evolution

Many of these are *not* outside the historic theory of neo-Darwinism, though some are. Rather, they stand in conflict with the modern version dominating textbooks and public thought.

Phenotypic plasticity

This issue arose earlier during the discussion of fossil teeth. The forms infant organisms grow into are critically influenced by the environment. The way our bodies develop is partially dependent on where we grow up (e.g., high altitude versus low altitude, equatorial latitude versus colder regions). Plants with identical genotypes will grow different leaves in one soil versus another. The temperature of incubation affects the gender of a gecko. Some aquatic invertebrates develop a carapace only in the presence of predators.

One effect of phenotypic plasticity is obvious: a population can change simply because its climate does. This change in form has nothing to do with genotype or natural selection. However, there is a more subtle way that phenotypic plasticity affects evolution. Phenotypic plasticity allows species to survive dramatic changes in environment without any shift in genetic makeup.

Even without mutations to help them, a species will change form to accommodate a new environment. Phenotypic plasticity allows a population to survive long enough for their genetic makeup to change. *More and more, biologists think the genes are not leading evolution but simply preserving it.* Who

A growing number of biologists see neo-Darwinism as explaining how changes are preserved rather than how they are caused.

knows how many significant changes in evolutionary history were only possible because phenotypic plasticity allowed a species to survive long enough for the genes to catch up?

Proponents of this view see genes as markers that lag behind the actual change. The DNA of a population may eventually come to encode the prevailing forms (so the phenotypic plasticity is no longer required), but it won't be *causing* the observed change in form; that has already taken place.

The importance of extinction

Many sequences of fossils in textbooks are misleading owing to phenotypic plasticity, but such sequences can be deceptive even for traits influenced by genes.

The fossil sequence showing horse evolution is infamous. Fossils can be arranged to suggest that horses once browsed on shrubs and later developed longer limbs and fewer toes, linearly evolving into the grazing horse of today. Some eminent paleontologists deride this presentation as misleading, claiming instead that many different kinds of horses developed with

A single catastrophic event can make irrelevant the gradual change over many thousands of previous generations.

no clear path of evolution. Several of these cousins were alive at the same time, and the one we have today is simply the single type that did not become extinct. Instead of a family tree depicting horses becoming gradually more and more adapted over time, Stephen Jay Gould casts it as an evolutionary bush showing a great variety of forms using myriad paths of adaptation.

This means that on a typical chart showing various fossils of horses many of the putative forefathers of the modern horse were likely not evolutionary ancestors but rather cousins several times removed belonging to a different evolutionary track altogether. The modern horse is not the summit of a long lineage of horses becoming more fit over time but rather the single remaining species among many viable cousins that lived contemporaneously. Whatever events caused the extinction of the other forms (which we wrongly consider forebears of the modern version) affected the history of horses more than all the eons of natural selection prior to it.

By way of analogy, consider a vast poker tournament at a bar. After several rounds of playing, the various gamblers have adjusted to each other, some with bigger piles of remaining cash than others. There may be some players that are doing much better than others, but only a few have actually had to leave the table. Then, one by one, they begin collapsing. Unbeknownst to the players, the barkeep had poisoned the beer. The players all eventually die except a single teetotaler who was neither the best or worst player there. The effect of the bartender was quite different from the effect of the poker. In evolutionary terms he caused a selection toward a trait (abstinence from drinking) far removed from the slower gradual selection of the poker game, which favored skill at cards.

Similarly, when we look back at fossils we may find a bunch of evolution that was adaptive to one environment but that ultimately had little effect on the overall evolution of a living species because some short-lived calamity selected the surviving species in a way that had nothing to do with the long-term environmental selection.

Genetic drift

The last example does not exhaust the ways in which momentous events can affect evolution. Anytime a catastrophe reduces a population to a very small size, the power of randomness increases. Sometimes even slightly harmful mutations present in such a group can gain a foothold and become standard to the descendant population. This power of chance, important in small populations, is called *genetic drift*.

Genetic drift can be critical in any situation involving small populations, not just catastrophic events. If a single pregnant rat and her mate are the only rats on a ship, and these stowaways take their leave on an island with no other rats, the genetic disposition of the ensuing population could be very different from that of the original population.

Gene duplication

Due to errors in the replication process, a long strip of DNA may occasionally be copied and re-added to a chromosome. This means that any genes on that length are *duplicated*. From an evolutionary standpoint, this is interesting because a cell only needs one copy to fulfill whatever functions a gene has.

Normally mutations are restrained because the original gene fulfills an important function. A mutant gene may be one step closer to coding for a very useful protein, but it probably no longer codes for the protein it once gave instructions for. Hence, the mutant dies or is severely disadvantaged. The one small step it represented toward a beneficial change is unlikely to go anywhere. However, if there are two copies of a gene, the mutation of one does not cripple the organism. Thus, one gene can be repeatedly changed in the background without leading to the immediate death most mutations cause.

In the schematics analogy, having a duplicate strip of DNA is like making a photocopy of a set of blueprints so that a colleague can freely modify it without risk of losing the original plans.

If the mutations that occur on this extra strand of DNA eventually allow the organism to code for a new protein that significantly changes it, the gap between parent and child may be so great that the new organism is not in direct competition with the older population. This means that natural selection cannot play its normal role because the species occupy different niches, like the mutant bird in the example described on page 39 whose diet differed from the main population.

Epigenetics

In the simple version of genetics taught to students, genes code for proteins and generally do not interfere with each other. This is very far removed from reality. There is a whole host of ways in which one gene can affect another; some of them are outright contradictions to Mendel's laws.[6]

In many cases the effect only lasts a single lifetime, but in June 2009 scientists cataloged over a hundred cases where the interference was heritable. Many of them resemble Lamarckian evolution,[7] which is regularly derided in science classes. Sometimes, interactions between an organism and its environment modify not only its own DNA but its offspring's as well.

In several cases, a person's behavior increases his or her vulnerability to a particular disease, and this vulnerability can be passed on. One example involves those who chew areca, a nut native to certain tropical regions. The children of areca chewers are more likely to have high blood pressure and diabetes than their peers, even if they never chew the nut. Researchers have

---

[6] One example is the gene regulation I mentioned earlier with the lactose-eating bacteria, where certain genes determine which other genes are active. In this section I'll discuss less artificial examples.

[7] Lamarck claimed that evolution occurred because the actions of one generation could influence the forms of the next. The oft-quoted example was that giraffes had long necks from many generations of stretching toward higher and higher fruit.

found that famine in one person's life can affect how their children develop, even if the famine had completely passed before the child was conceived.

It's hard to see how these particular examples could wield any material influence over the whole course of evolution, but they do indicate that there

> The symbiogenesis theory claims early organisms evolved through networking rather than through competition.

is more to heritable genetics than we may know at this time. Epigenetics is a young field, and its deepest mysteries have almost certainly not been plumbed. Even now, some experts in the field believe it shows evolutionary biology took a completely wrong turn with Darwin. Søren Løvtrup has written several books on the subject and claims "Darwinism is the greatest deceit in the history of science."

Horizontal gene transfer

Horizontal gene transfer is a fancy way of referring to an organism's absorbing or incorporating genetic material from a non-parent, either from its own species or another. Obviously this is well outside the bounds of how we generally think of heredity. The new DNA can cause vast changes in the offspring, far beyond the reach of natural selection's influence.

Moreover, the engulfed DNA can *itself* mutate and then get passed around to other organisms. This is how a population of bacteria "learns" to digest nylon. The critical mutation does not occur in the bacteria's germ DNA but in the DNA of the plasmids they contain. Instead of one bacteria evolving to digest nylon and that DNA's progeny overwhelming the rest (survival of the fittest), mutated plasmids that tweak already-existing enzymes get passed around to spread the adaptation.

A far more important example of horizontal gene transfer has been proposed to explain the acquisition of mitochondria by eukaryotes. Mitochondria are the organelles that provide energy to our cells. The accepted theory for how we came to have them is that they are the evolved progeny of symbiotic creatures who happened to get stuck inside a cell long, long ago. They then replicated separately and have stayed with us ever since. A similar hypothesis is proposed for the evolution of chloroplasts and flagella.

This theory's greatest advocate has been Lynn Margulis, who began showing evidence for it in 1967. Of course, the idea that such a huge shift in evolution could occur in a single step through pathways having nothing to do with the established evolutionary theory struck many biologists as ludicrous, and her colleagues hurled a great deal of invective at her for even suggesting it. The feeling was mutual; Margulis famously declared that history will ultimately judge neo-Darwinism a "minor twentieth-century religious sect within the sprawling religious persuasion of Angle-Saxon biology."

Her views on neo-Darwinism aside, the intervening years have found biologists growing more and more supportive of her theory in the case of chloroplasts and mitochondria. It is now accepted by most evolutionary biologists. Recently, it appears a slug has managed to become photosynthetic by incorporating DNA from an algae, and an amoeba, *Paulinella chromatophora*, has done the same.

Margulis hypothesizes that the nuclear membrane itself is due to such an absorption of one organism into another. Philip Bell proposes a more radical notion: that eukaryotes originated by a viral infection of a prokaryote.

Clearly, in the overall scheme of evolution, these steps are massive. Furthermore, it appears that horizontal gene transfer is much more common in prokaryotes, so its influence during the earliest evolutionary phase could be even greater than afterward. We may come to find the overall effect of these huge jumps dwarfs those of mechanisms familiar to neo-Darwinism.

A Turning Tide?

Modern evolutionary research on the mechanisms discussed above has paralleled a growing sense that the evolution we teach in school is not sufficiently compatible with the fossil record. Genetic mutations appear too mild, too inefficient, or too intractable to account for the program shown in the fossil record, where a vast amount of biological *disparity* appears to have developed in a relatively short timespan long, long ago followed by no significant new structures afterward.[8]

Evolutionary biologists have responded in diverse ways. A few, like Richard Dawkins and David S. Wilson, have patched the theory with ancillary ideas. Several scientists find these modifications sufficient to preserve the traditional view's plausibility. Others, like Gould and Niles Eldredge, have attempted to drastically alter the theory's emphasis and perspective without abandoning it. Others have searched for new, radical ideas that would require the theory to be essentially decommissioned. One is *directed mutation*, the belief that mutations are not random but can be influenced by the environment. Another is Margulis' symbiogenesis. Several scientists have attempted to revive Lamarckism, including Pierre-Paul Grassé, the legendary French zoologist responsible for the 38-volume *Traité de Zoologie*.

Connected to this movement is a growing belief that the type of evolution taught in school is good for stabilizing populations after massive change has occurred but not adequate to explain how such change came about. To quote Scott Gilbert, "the modern synthesis is remarkably good at modeling the survival of the fittest, but not good at modeling the arrival of the fittest." Paleobiologist Graham Budd puts it more bluntly, "When the public thinks of evolution, they think about the origin of wings and the invasion of the land. But these are things that evolutionary theory has told us little about."

---

[8]I discuss this issue more in the third volume's final chapter.

Conclusion

Modern teaching of evolution suffers from two significant problems. First, the limited, artificial version of Mendelian genetics presented to students does not properly support the neo-Darwinian perspective. Efforts to transition from genetics to evolution are practically doomed from the beginning, and textbooks cope by glossing over the problems and concerns that challenged scientists for forty years. They try to sell evolution to students by presenting to them a bargain that the best scientific minds of the early 20th century would not buy.

Secondly, modern research has exposed several issues problematic to traditional neo-Darwinism. They include both a collection of evolutionary mechanisms lying outside the common articulation of the model and a growing sense that the modern synthesis does not adequately explain the change we see. The question is "What mechanisms are to be credited with most of the evolution experienced during the course of Earth's history?" The answer is becoming less and less clear.

But what should students be told until then? I would suggest giving them an honest description of the current state of evolutionary thought. Something like:

Genetic mutations and natural selection, once thought to dominate the history of life's evolution on Earth, have proven ill-suited to predict, or even explain, the changes we see in the fossil record. New mechanisms for evolution, or new interpretations for established ones, have been suggested to remedy this inadequacy. While some biologists are satisfied with these modifications to the theory, a growing number feel a different perspective altogether is welcome and necessary. It is likely the prevailing view will shift to a model where genetic changes and natural selection take a less and less prominent role as one looks further and further back into history.

Why Are Veins Blue?

Pale skin bears witness to a curious fact: veins trace a blue path across the forearm even though we all know blood is quite red. Several false explanations are given for this. The most common myth runs something like:

> Blood's color depends on whether it contains oxygen. Most of the blood we see in everyday life is in contact with oxygen, which turns it red. Our veins, however, carry oxygen-free blood back to the heart to be pumped to our lungs. Lacking oxygen, this blood is blue instead, giving veins their characteristic hue.

Teachers are understandably tempted to give this explanation for the blueness of veins; it shows students a visible reminder that veins[1] carry blood depleted of oxygen, having already distributed it to cells around the body. Still, it's odd the myth persists, since millions donate blood (taken from a vein) every year and see their gift of life has a very non-blue hue. If the myth about blood were true, we would expect blood flowing through a plastic hospital tube to be blue, except possibly at the leading edge of the flow. The rest of the blood in the tube would be blocked away from whatever scant oxygen had been in the tube before the blood was drawn.

In fact, blood from the veins (called *venous blood*) is a dull, dark red—no blue whatsoever. Nor is this blood oxygen free. It has significantly less dissolved oxygen than it had when it left the heart, but much more than most other liquids encountered in everyday life. Venous blood carries about twenty times more dissolved oxygen than water can hold under normal conditions.

---

[1]Throughout this chapter, I refer to veins near the skin of fair-skinned adults. Blood carried by the *pulmonary vein* is leaving the lungs and is loaded with oxygen, but this vein is hidden by the sternum. The umbilical vein also carries oxygen-rich blood, but I assume most readers lost theirs years ago.

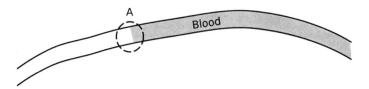

Other than blood near the edge (region A), the blood drawn through a hospital tube is not in contact with atmospheric oxygen.

The myth gets one detail right: oxygen-rich blood is bright red. *Hemoglobin* molecules are responsible for carrying oxygen in human blood. They change shape when picking up an oxygen molecule, and this turns them from purplish blue to bright red.

Beyond this, the myth falters. Though the deoxygenated form of hemoglobin is dark blue, these molecules are outnumbered by their red counterparts even in venous blood, typically by a margin of 3 to 2. Moreover, even if blood were dark blue, it wouldn't make a material difference in the vein's appearance. After all, venous blood is a dull, dark red, yet we do not see cherry chocolate lines running along pale forearms.

Those sources that do not succumb to the "blue blood" explanation often indicate that veins appear blue because only blue light penetrates far enough to hit the vein and bounce back to be seen. Other colors of light, the story goes, are absorbed by fat beneath the skin, never reaching our eyes.

There are a couple of problems with this "penetrating blue light" theory, but the most obvious is that it suggests veins are the only things reflecting light. This cannot be true; many people have skin much lighter than their veins, which would be impossible if the fat and pigments beneath the skin were such avid absorbers of red, green, and yellow light. We will find, ironically, that veins appear blue because blue light is *less* likely to penetrate deeply!

The Truth Is Skin Deep

Plastic wrap is *transparent*; light can pass through without much blurring. Wax paper is *translucent*; some light passes through, but it gets scattered by the molecules in the paper. The more layers of wax paper you put together, the less light makes it all the way through before being absorbed. Skin low in *melanin*, a family of pigments, is like wax paper in this regard.

Very little of the light hitting the skin manages to penetrate down to the vein, bounce off it, and then escape to be seen by our eyes. This means that when pale-skinned people look at their veins, they are really seeing the fatty tissue lying beneath the skin's surface.[2] Some of that tissue has a dark background (the vein) behind it, and some does not. It's the difference between how wax paper looks when it's on top of more wax paper versus how it looks when it is covering a chocolate bar.

The short answer to why tissue above the vein looks blue:

1. Red light is more likely to penetrate to the vein (and be absorbed) than blue light, so less red light escapes the skin above a vein than it does veinless skin.

2. The contrast in the amount of red light escaping veinless skin compared with skin above a vein tricks our brain into seeing the latter as bluish, even though the light coming from that region is not actually blue.

The rest of this chapter is devoted to investigating these statements. Throughout the discussion we compare red light with blue light because they are on opposite ends of the spectrum. Green light is in the middle of the spectrum, and our skin treats it similar to blue light. Hence, we can draw rough conclusions about the overall effect of a vein on the whole spectrum by investigating just red and blue light.

---

[2] Dark-skinned people have high levels of melanin in their lower epidermal layers. These pigments absorb most visible light, making the skin dark.

1. Wax paper skin

When light hits fair skin, the fatty tissue in the hypodermis kicks it around like a ball in a pinball machine. This scattering ejects some of the light from the skin; the rest is absorbed. How we see the skin depends on the light that escapes, for only such fugitive light makes it into our eyes. A dark background, like a vein, changes the skin's appearance by absorbing light before it can escape.

> The color of the blood in our veins is largely irrelevant. What matters is that it is dark.

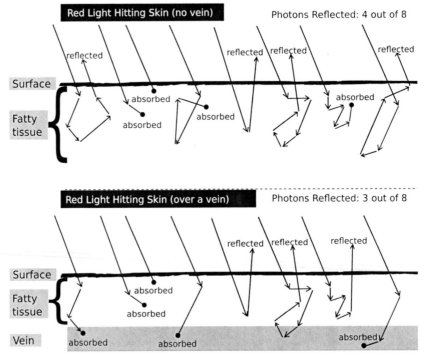

Veins reduce the amount of light escaping the skin above them.

If the vein simply absorbed some of the light rattling around in the skin, it would only darken the skin, not alter its color. The particular chemicals in fatty tissue absorb red less easily than blue or green. This means red light penetrates farther than blue or green, so most of the light absorbed by the vein is red.

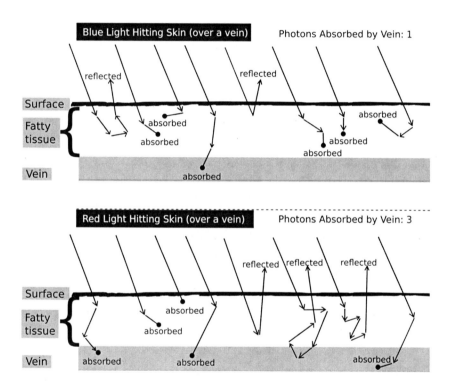

Blood in a vein absorbs more red light than blue simply because most blue light never reaches it.

A key point to take from the previous two pictures is that light hitting fair skin can be classified into three categories:

1. Much of the light is absorbed by the skin and does not affect the color we see.

2. A large amount of the remaining light is reflected by the fat below the skin after bouncing around a bit inside.

3. A small amount of light hits the vein. Most of this is absorbed, but some red light is reflected back.

As long as the vein is deep enough, blue light tends to split roughly evenly between the first two categories while red light is more evenly distributed among all three.

Even though low-oxygen blood will reflect a small amount of red light, most of it is still absorbed, meaning that the vein decreases the amount of red light bouncing back up to our eyes more than it decreases the amount of blue light. Blue light is less affected because it tends to be either absorbed in the skin or reflected back upwards without ever reaching the vein.

Thus, venous blood paradoxically *lessens* the amount of red light we see coming from the skin above it. This has nothing to do with the color of the blood. If the blood were dark green, dark blue, dark purple, or any other dark color, it would have the same effect. What matters is that the blood in the veins absorbs most of the light that hits it; because pale skin is chemically configured to allow red light to penetrate farther, most of the light that ends up being absorbed is red.

Things would be different if veins were closer to the skin. The less tissue between the skin and veins, the more light reaches them to be absorbed. If veins were very close to the skin, then almost all the light would reach the vein, and the slight reddish hue of the blood would become a factor. Instead of the vein absorbing more red than blue (because more red hits it), it would absorb less red than blue since it naturally reflects red better than other colors.

> Veins lessen the amount of light escaping fair skin. Red light is affected most, for it penetrates deepest.

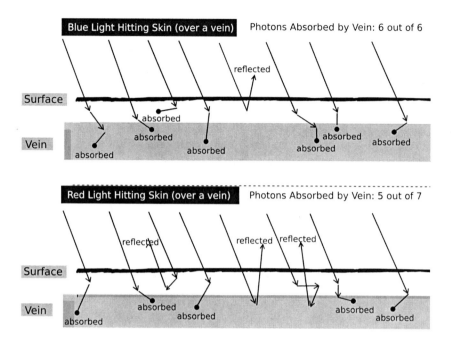

If a large vein were very close to the skin, it would look red because the slight reflection of red light from the blood in the vein would be a greater factor than the shallower penetration of blue light.

This scenario does not typically occur because we don't have veins close enough to the skin that are large enough to absorb much light. Still, it is a useful picture to contemplate as it shows there are two factors at work. Blood, even oxygen-poor blood, naturally absorbs blue light better than red light. That is why it has a reddish color. However, since red light penetrates more deeply, there is more red light hitting the blood. Hence, the question of whether the blood is absorbing more red light or more blue light depends on the depth of the vein.

The appearance of a vein depends on its depth.

2. Color confusion

I've presented several illustrations showing why venous blood tends to absorb more red light than blue light as long as the vein is suitably deep. One might naturally wonder, though, why this is the least bit relevant! We normally think the color of an object is attributable not to the colors it is absorbing but the colors it is reflecting. In the figure comparing blue light with red light on page 53, the skin reflected more red light (3 photons out of 8) than blue light (2 photons out of 8). Shouldn't this mean the skin appears more red, regardless of all the discussion about which light the blood was absorbing?

In short, no. Our brain does strange things when it processes the light received from our eyes. The color it perceives depends not only on the light coming from an object but also on the light coming from its surroundings. *Contrasts* between colors affect our perception, so a shirt may look more orange or less orange when seen against different backgrounds. When considering the color of skin above the veins, the contrast between it and the surrounding skin turns out to be more important than the actual color ratios. If the skin above the vein reflects about 2 out of every 8 blue (or green) photons and 3 out of every 8 red photons, we would expect to see a dark reddish-gray color—a mauve taupe. Our brain overrules the objective color combination and registers the vein as bluish instead.[3]

One could design an experiment where a small hole is poked in a shirt and maneuvered so that only a small section of a vein region is visible. Close-up pictures could be made and given to people who do not know they are looking at a hole in a shirt. This would remove our psychological conditioning to see veins as blue. It would be interesting to see how people identified the color shown and to what degree it depended on the shirt's color.

---

[3]A branch of optics known as *retinex theory* has been developed to factor in the background color.

Percent of Light Escaping Skin

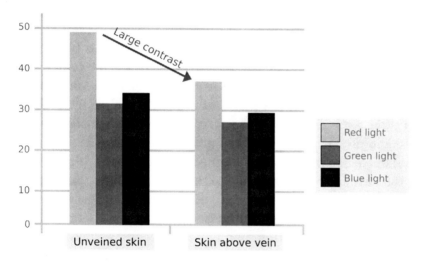

The presence of a vein sharply lowers the amount of red light seen coming from the skin above it without having as significant an effect on light of other colors.

Conclusion

The blood in our veins is not blue. More surprisingly, there is more red light coming from the region above the vein of fair-skinned people than any other color.

The reason veins appear blue is that the blood inside them absorbs most of the light that penetrates deeply enough to hit them, and the pigmentation of fair skin allows red light to penetrate further than light of shorter wavelength (e.g., blue, green). The blood in our veins absorbs much of the red light striking the skin above them, creating a significant contrast between the region immediately above the vein and the adjacent skin. This contrast tricks our brain into seeing blue even though our eyes still see more red light.

Addendum: Why is the Sky Blue?

After all the above discussion of how different colors of light are scattered or absorbed at different rates, I cannot refrain from giving a quick explanation as to why the sky is blue.

When we look at the sky, most of the light we see comes from the Sun. Of course, if the Sun is near the eastern horizon and you look straight up, the light you see could not have traveled a straight path. Sunlight is scattered by the atmosphere, zigging and zagging all about (just as light scatters about in your skin) so that we see light coming from all over the place. If the Sun is in the east and we look up and to the west, the light that hits our eyes must have done a significant amount of course changing to get there. It is easiest for light with short wavelengths (blue, violet, indigo) to scatter than colors with longer wavelengths (green, yellow, red). This means blue, violet, and indigo light waves have an easier time making the circuitous route required so that they appear to come from a patch of empty sky. Even though violet and indigo light scatter more easily than blue, the sky appears azure because our eyes are more sensitive to blue and the Sun emits more blue light than indigo and significantly less violet light than either one.

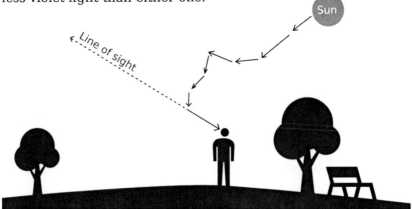

Producers and Consumers

Most of the problems described in this book refer to faulty descriptions or misconceived interpretations of real-world phenomena. When discussing the classification of organisms in an ecosystem, there looms a more fundamental issue: textbooks provide definitions that manage to be incorrect, ambiguous, and misleading all at once. I don't recall ever seeing a

"Producer" and "consumer" do not describe where an organism gets its energy, but rather where it gets its carbon.

textbook for middle- or high-school students give the correct definitions for *producer* and *consumer.*

When students learn about food chains and predator-prey relationships, they are taught to focus on the question, "Where does the organism get its energy?" This naturally arises as they study respiration and photosynthesis. Children are told that plants are producers because they *get their energy from the Sun.* They are told other organisms are consumers because they *get their energy by eating plants or each other.* A popular alternative is to replace "energy" with "food," but we will see that this is also untenable.

The terms "producer" and "consumer" are a bit plebeian for scientists. Instead, they use the terms *autotroph* (for producers) and *heterotroph* (for consumers). The correct definitions for these terms do not refer to where an organism gets its energy, but rather where it gets its carbon. The definitions given in school do not merely oversimplify, they mix-and-match scientific vocabulary. Scientists do have a term for "an organism that relies on organic matter for its fuel." Our textbooks call such an organism a heterotroph, but it is actually the definition of an *organotroph.* Not all heterotrophs are organotrophs; not all organotrophs are heterotrophs. So, the question of what fuels[1] an organism is an important one, but it isn't the most important, and it is not the defining question determining whether something is a producer or not.

The downside of this error is not limited to incorrect definitions and mixed-up vocabulary. Textbooks presenting this classification system wrong students in three other ways:

- These definitions squander a prime opportunity to inform students of the fundamental importance of carbon. One specific infelicity is the pigeon-holing of photosynthesis as merely an expedient for acquiring energy.

- The two options presented (energy from sun versus energy from eating other organisms) do not exhaust all options, leaving students to guess at how to classify creatures getting their energy from another source.

- Behind these definitions is the idea that producers are self-reliant, at least in terms of their nutritional needs, and form the base of an ecosystem. This is a great fiction that skews how children think about living systems.

Carbon is the Key

All life with which we are familiar is based on carbon. Carbon's abundance and ability to form up to four bonds with nearby atoms make it nature's perfect choice for supporting the large molecules found in every living cell. It is not surprising, then, that all growth, development, and repair of cells require carbon. Scientists classify organisms as autotrophs or heterotrophs based on what forms of carbon they can use.

---

[1] I use "fuel" here instead of "energy" because it better matches what textbooks actually mean. See *Notes* chapter for a discussion on all three classifications, by source of carbon, fuel, or energy.

Carbon atoms are so prevalent in organic compounds that chemists typically omit them when drawing molecules. It is understood that any unlabeled vertex represents a carbon atom. Hydrogen atoms attached to carbon are often not even drawn at all. Scientists can infer where they are without vertices representing them. The left diagram shows how an organic chemist might draw a glucose molecule. The figure on the right shows the same molecule with all atoms labeled.

A *producer* is an organism that can survive with carbon dioxide as its only source of carbon. Creatures lacking this ability are consumers.

The above definition is not meant to imply that producers never use anything else for their carbon needs. Many use a combination of sources in the wild. However, to be an authentic producer, an organism must *have the ability* to synthesize all the long, complicated compounds necessary for life without these other sources. They can produce *organic* material from *inorganic* building blocks.

Students are taught to see photosynthesis as a harvesting mechanism: plants package solar energy into sugars. Our fixation upon energy shows through in the equations given in biology textbooks, which portray glucose as the product of photosynthesis. In reality, very little glucose is produced. Chloroplasts, the organelles utilizing photosynthesis, are essentially microscopic spinning wheels producing threads of carbon rings to be used in

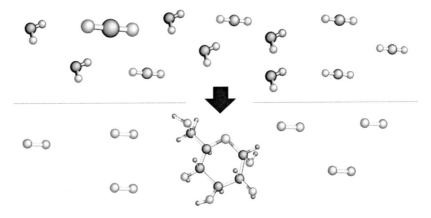

Plants are producers because they are able to synthesize the complex organic compounds required for growth and repair from simple inorganic molecules. The figure above shows six molecules of water (the bent ones) and six molecules of carbon dioxide (CO_2) being melded into a single glucose molecule ($C_6H_{12}O_6$) and 6 molecules of oxygen gas (O_2).

many different ways. Saying "photosynthesis produces glucose" is like saying bakers produce doughnut holes.

A carbon-based perspective asks them to see it in a different way. Instead of emphasizing what photosynthesis does to energy, they are asked to consider the reaction's effect on matter. Green plants synthesize complex compounds by gluing simple CO_2 particles together at the molecular level and sprinkling on some hydrogen ripped away from good ol' H_2O. Most plants take in carbon dioxide and use it for growth and repair. Practically all of a tree's weight comes from molecules that were originally in the air. The plant is literally building its own branches and leaves with repurposed carbon dioxide!

Plants provide ecosystems with a service even more fundamental than capturing energy from the sun. Plants and other producers are amazing nano-factories that build up complex, organic compounds from tiny inorganic molecules. They glue back together what the rigors of life break apart. This process is so central to ecology that it is called *primary production*. No wonder the correct ecological definition for "producer" refers to how carbon is treated rather than how energy is obtained.

A Wide Middle Ground

Nature's wonderful variety presents challenges to students trying to categorize organisms on the basis of how they acquire energy. Hydrothermal vents in the ocean floor glow enough that some bacteria can conduct photosynthesis, and black mold at Chernobyl converted the gamma rays there into usable energy. My guess is that most bright students could see the similarity between these organisms and typical plants; in both cases energy is coming from an immaterial source. But what about organisms that don't feed on others but get energy by processing matter?

Are the bacteria that get energy from "burning" ammonia producers? What about bacteria obtaining their energy by converting carbon dioxide to methane? What about bacteria that only need methane? What about mushrooms? Earthworms? Dung beetles? Vultures? The sources of energy (either at the molecular or macroscopic level) for each of these creatures is more complex than the last. A student could easily infer that none of them are producers and then wonder why her teacher says the first two are producers while the last five are not. An energy-centric definition for these terms not only requires students to guess at how to classify certain organisms, but it also often fails to hint at the correct answer. Students armed with the definitions supplied by our

> Producers glue together what the rigors of life break apart. This process is so central to ecology that it is called "primary production."

textbooks have no hope of understanding why a bacterium that gets energy by "eating" carbon dioxide is an autotroph while one that "eats" methane is not.

These in-between feeders, whose energy comes from terrestrial, non-living sources, are not the only ambiguous cases. Consider, for example, algae that utilize photosynthesis but live in environments requiring them to get energy from other sources as well. The reverse also occurs, some organisms (like purple non-sulfur bacteria) get energy from the Sun by photosynthesis but cannot build up the necessary organic compounds for life.[2]

These examples highlight a nuance behind how "producer" and "consumer" are commonly discussed in textbooks. Their descriptions ascribe more importance to an organism's dependence on others than to its contribution to an ecosystem. Any philosophy that identifies a population's role with what it relies upon is prone to many problems, especially when it incorrectly suggests plants are self-sufficient.

Self-Reliance is Mostly Illusory

The definitions presented by textbooks promote a conceptual schema where consumers are viewed as a broad class of moochers depending on producers, who are hailed as self-reliant. This interest on *dependency relationships* can corrupt our analysis of ecological roles because it distracts us from focusing on a population's contribution to the ecosystem at large.

To see what I mean, consider methane-eating bacterium populations in two different ecosystems. The methane in the first ecosystem is created naturally by physical processes. For

---

[2]You might wonder how it is possible for photosynthesis to occur without organic molecules being formed. The short answer is that there are different kinds of photosynthesis, and some creatures only use the energy-trapping, ATP-forming reactions (the "light reactions") and not the sugar-creating steps (the flagrantly misnamed "dark reactions," which do not occur in the dark.)

example, it may be ejected from the mantle through cracks in the ocean floor. In the second ecosystem, the methane is waste produced by other organisms. A philosophy that emphasizes dependency leads one to consider the methane-eaters in the first ecosystem as producers

> Practically every population in every ecosystem relies on other populations.

but views the identical species as a consumer in the second. Rather than focus on how the methane-eaters contribute to the ecosystems (which is the same in both cases), it defines their role in terms of their reliance on other organisms. This is a fool's game because practically *every* population in every ecosystem relies on other populations, which is why the alternative definition using "food" instead of energy is not a useful one. "Food," properly understood, refers to any form of nutrition required for growth and development. Contrary to what is often presented, plants are independent in this regard.

Carnivorous plants, like the Venus flytrap, do not get energy from eating insects, but they do rely on them for nutrients. In fact, this situation where a plant depends on other organisms for nutrition is very common. More than 90% of all plants are in a mutual feeding relationship with fungi living on their roots. The plant feeds on the fungi's cells, taking in raw nutrients they have absorbed from the soil while the fungi absorb complex organic molecules the plant assembles. It is worth noting that these associations do not merely increase the total amount of nutrition the plant can take in. The fungi not only absorb material better from the ground, they also break down organic compounds into simpler components. Scientists have deduced that plants fundamentally lack the physiological ability to do this.

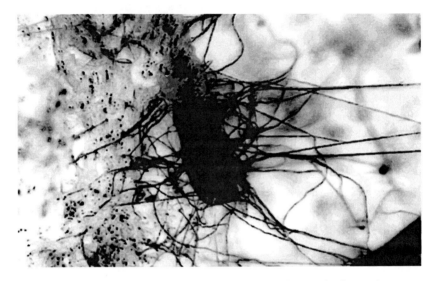

Most plants, though technically producers, effectively depend on my-corrhizae, symbiotic associations with fungi growing on their roots. The fungi, the strands in the picture above, provide nutrients required for photosynthesis and other processes. The large mass in the middle is a piece of the plant's root. The stringy fungi can absorb nutrients from the soil much more effectively than the root. They leech organic compounds the plant produces. Plants with limited or no ability to perform photosynthesis can cheat in this relationship, taking both organic and inorganic nutrition from the fungi without giving anything back. Photo courtesy of Paula Flynn, Iowa State University Extension.

The Diversity of Bacteria

Microorganisms offer lots of examples illustrating why the energy-centric definition given in textbooks could lead someone to misclassify an organism.

Heliobacteria get all the energy they need from the Sun, but they still require organic carbon, which they likely absorb from rice-plant secretions. Since they require a source of carbon other than CO_2, they do not qualify as producers. Other microbes, labeled *organoautotrophs*, get their carbon from carbon dioxide, but require organic matter for metabolism. Even though these microbes rely on organic matter, the organisms are considered autotrophs (producers) because they are able to use inorganic compounds to build up proteins and other growth components.

Then there are the aforementioned methane-eating bacteria. They cannot be considered producers because methane is an organic molecule. However, these consumers can satisfy the carbon needs of an ecosystem. This gives another example of how the typical presentation of producers as the base of an ecosystem is seriously flawed.

Similar to green plants, which are producers but rely on mycorrhizae for their nutrition, bacteria also exhibit mutual feeding relationships. Half of a giant tube worm's body weight is typically composed of bacteria that assemble the nutrients it absorbs from the ocean. The bacteria feed on the worm's blood (their only source of hydrogen sulfide) while the worm absorbs the sugars built by the bacteria.

Conclusion

Life is extremely complex, and depends on a vast array of chemical processes involving a multitude of reactants. Plants cannot survive on just light, carbon dioxide, and water. The process of photosynthesis requires many other critical components, and the green plants considered the basis for an ecosystem generally rely on other organisms directly or indirectly. Most plants depend on other creatures in a practical sense, as in the mutual feeding relationship of plants and fungi. If being a producer meant not feeding on other creatures, most common photosynthetic plants would fail to make the cut.

Producers are so called based on their ability to synthesize required organic compounds from simple inorganic molecules, even if they rely on other organisms for food (nutrition) and/or energy. Conversely, if an organism cannot synthesize organic matter from carbon dioxide, it is not a producer, even if it obtains energy from the Sun or another non-living source. The classification is based on a population's' contribution to the ecosystem, not on its reliance upon other populations.

If being a producer meant not feeding on other creatures, most common photosynthetic plants would fail to make the cut.

What should we teach children? Once they are in middle school, the terms "organic" and "inorganic" can be defined, allowing a proper treatment of producers versus consumers. At the elementary level, we can convey the essence of the correct definition without using technical jargon.

While the urge to introduce the energy-trapping ability of chlorophyll is hard to fight, I would suggest we simply tell younger students that life depends on many very complicated substances. Some creatures have the ability to construct these substances from simpler ones. Everything else is a consumer, and decomposers are special consumers that break down complicated substances into simple building blocks the producers can use.

Few topics suffer such a variety of explanations as the tide on the "far side" of Earth, the side farthest from the Moon.

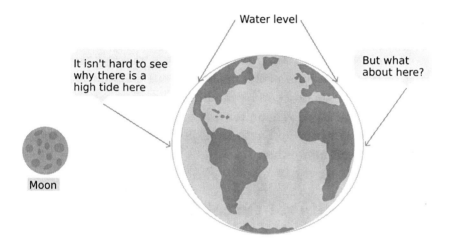

Tides in a given location are influenced by several factors. Some coasts have one high tide a day; others have several; most have two. This chapter discusses the standard, two-a-day tides most locations experience.

One high tide lags a little behind the Moon and is easily (though incompletely) explained by saying the Moon's gravity pulls the water toward it. The other high tide is then harder to understand. If the Moon draws water toward it, shouldn't the other side of the planet be at low tide?

The most common explanation given to students is that this "far tide" is due to *centrifugal force*, the sensation you experience when making a high-speed turn while driving. You feel your body thrown a bit toward the outside of the turn. For similar reasons, students are told, water is thrown outward (away from Earth) as our planet curves through space.

An alternate explanation is: *The Moon pulls the land toward it, leaving the ocean behind, so those on the shore see the water rise relative to them.*

In this chapter we investigate these explanations and their deficiencies. The first cannot be a satisfactory description since "centrifugal force" is not a real force. It cannot be the actual cause of anything. The second regrettably attributes a useful observation to the wrong cause, leading to confusion. After briefly describing the problems with these explanations, I will show how the second can be modified, clarified, and extended to explain the basic reason for tides. We will also see how the prevalence of water on Earth amplifies our tides in a way most textbooks omit altogether.

The "Centrifugal Force" Explanation

We generally think of the Moon as orbiting Earth, but in reality both orbit another point, their *center of mass*, which is inside Earth—closer to the surface than the center. Think of spinning a young cousin through the air in a circle. You have to redistribute your mass as you spin to counterbalance your cousin's moving mass. You wobble in a small circle while your airborne relative flies around you, clutching for dear life. Similarly, Earth wobbles in a small circle about[1] the center of mass as the Moon moves in a large circle around the same point (which moves around the Sun). Earth and the Moon are essentially a mismatched couple dancing a Viennese Waltz on a cosmic scale. It is *this* circular/elliptical motion people refer to when describing how "centrifugal force" causes tides.

Two problems arise when citing centrifugal force as the cause for tides. First, "centrifugal force" is not real. When

---

[1] It is common to use "about" to refer to motion relative to a point interior to an object and "around" to refer to motion relative to a point external to it. So, Earth rotates *about* its axis but revolves *around* the Sun.

you make a sharp turn while driving, your torso is not thrown outward. Rather, the seat of your pants is pulled inward owing to friction. Your car is try-ing to go in a new direc-tion, and your body is still going in the old one. The car is in contact with your body, so it pushes your

"Centrifugal force" is not a real force. It cannot be the actual cause of any phenomenon.

body in the new direction. The forces (and your body's reaction to them) give the impression of an outward force that does not really exist.

Therefore, it is a safe bet that a middle-school text attributing a phenomenon to cen-trifugal force has taken the wrong path. Physi-cists sometimes use it as an artificial, math-ematical tool for cal-culations, but centrifu-gal force should never be portrayed as the ac-tual cause for anything. Some books use "force of inertia" instead of "centrifugal force." In-ertia is real, but it is not a force and can-not be the fundamental cause of tides.

While simply say-ing "centrifugal force is not a real force, so it

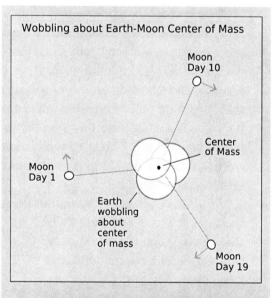

Wobbling about Earth-Moon Center of Mass

Moon
Day 10

Center
of Mass

Moon
Day 1

Earth
wobbling
about
center
of mass

Moon
Day 19

The Moon and Earth orbit their center of mass. This causes Earth to wobble in a circle over the course of a lunar cycle. Many textbooks incorrectly claim tides are caused by "centrifugal force" due to this motion.

cannot be the cause of tides" is a perfectly valid objection, there is a second concern: we would have tides even if the Moon and Earth did not orbit their center of mass! Indeed, we will find tides easier to understand if we pretend their cosmic Viennese Waltz were stopped.

The "Pulling Down the Shore" Explanation

Sometimes tides are explained by saying that gravity pulls the ground toward it, leaving the ocean behind. According to this explanation, the people on the shore see water rise, but in reality it is they who are being pulled downward.

The mental picture evoked comes close to capturing a key idea, but the actual statement is a queer explanation indeed— it suggests gravity affects only land! A student taught this reasoning is in a lamentable situation, for she then has to wonder what causes the tide on the shore *nearest* the Moon. The idea that the Moon pulls the water toward it was a simple explanation for *that* tide, but now the student is led to believe the Moon pulls only on land. Unless gravity is very selective, affecting only water on the near side and only land on the far side, the whole affair seems hopeless.

Students may also believe that there is an actual gap between the land at the ocean floor (which has been "pulled away") and the ocean. Or, if they realize that the water moves with the land, then they are confused because some other chapter of the very same textbook may have told them that liquids have a *definite volume*. If the land is pulled away from the ocean, and there is no gap between them, then the water had to expand.

> Gravity does pull the shore downward, but it pulls the water downward as well.

The Causes of Tides

Orbital motion in the Earth-Moon system is just a distraction. It is much easier to understand tides if we ignore it.

The Moon constantly "falls over" Earth. Gravity pulls the Moon toward Earth, but the Moon is going so fast sideways to this pull that it perpetually misses, keeping the

The forces causing tides on Earth would cause them even if the Moon and Earth were not circling their center of mass.

Moon in orbit. Were this motion halted, the Moon would fall directly to Earth with unwelcome consequences.

Pretend super-villains leap from the pages of a comic book and into our solar system to bring about those "unwelcome consequences" by halting the Moon's sideways motion. Just to be sure the collision is dead-on, let's say the super-villains stop Earth's wobbling motion as well.

If the super-villains stopped the sideways motions of the Moon and Earth, gravity would pull them towards one another. We would have between four and five days to make peace with the Creator prior to the crater. Tides would occur on both sides of Earth during that time, even though there is no longer any orbital motion in the Earth-Moon system. (See diagram overleaf.)

Tides form in this scenario for the same reasons they occur in real life. The forces acting on Earth in the second frame of the diagram are identical to the forces acting in our everyday existence. Orbital motion does not cause tides, it just lets us *observe them safely* as the Moon falls *over* Earth rather than toward it. Tides would occur even if the Moon and Earth were at rest and able to approach one another directly. But surfers would not have long to enjoy them since the Moon would impact Earth in less than a week.

What If Super-Villains Stopped the Spinning?

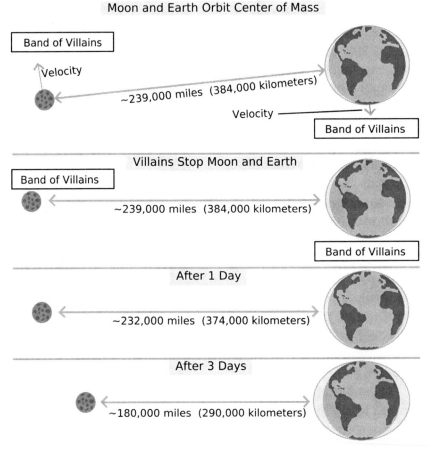

Moon and Earth Orbit Center of Mass

Band of Villains

Velocity

~239,000 miles (384,000 kilometers)

Velocity

Band of Villains

Villains Stop Moon and Earth

Band of Villains

~239,000 miles (384,000 kilometers)

Band of Villains

After 1 Day

~232,000 miles (374,000 kilometers)

After 3 Days

~180,000 miles (290,000 kilometers)

Even if the Moon and Earth were stopped in their tracks by bands of villains, removing Earth's circular motion, gravity would still cause tides on both sides of Earth. It would take nearly five days for the Moon to finish falling toward Earth. (The Earth would move a tiny bit toward the much lighter Moon—too little to show here.)

The tides on Earth are primarily due to four factors:

1. The gravitational force between Earth and the Moon

2. Differences in structural forces between solids and liquids

3. A consequence of the Pythagorean theorem

4. The tendency of water to flow

Factors 1 and 2 combine to form the basic high-tide effect expected on any planet, a consequence of the Pythagorean theorem causes low tides to be lower than we might expect otherwise, and water's fluidity accentuates both low and high tides on Earth.

Basic high tide effect

Some books make an honest effort at explaining tides by referring only to gravitational force. For example, they might claim that gravity pulls the ocean floor (and the rocky shore) away from the water. While this explanation has problems, it's on the right track. The shore is being pulled toward the water, but not by gravity. Gravity, after all, pulls equally strongly on both the land and the water.

The problem with these explanations is that they don't consider how matter *responds* to the gravitational force. Indeed, these explanations imply a naïve conception of gravity's effect that would require a *low tide* on the shore opposite the Moon: lunar gravity pulls less on the shore than it does on the water's surface—the shore is higher than the surrounding water and hence farther from the Moon.

A satisfactory explanation of tides should include not only gravitational force, but also the reaction to its pull by the water and land.

75

Gravitational Stretching on Shore Farthest from Moon
(Ignoring Structural Forces)

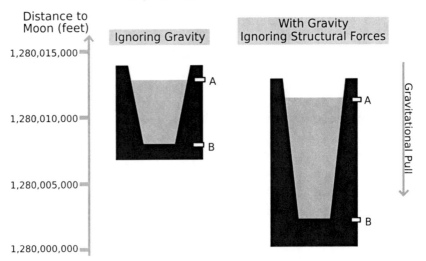

The figure shows the effect of the Moon's gravity on the shore farthest from it. Material closest to the Moon gets pulled downward more than material farther away. This means the ocean floor is pulled farther down than the water and land at the surface. The figure shows what we might see if matter could freely expand. (This figure is not to scale and is exaggerated for illustration.)

But, of course, this doesn't happen. We do not see low tides there. Why does the water level *rise* compared with the shore when it is closer to the Moon (on the opposite side of Earth)? The Moon pulls most strongly on material that is closest to it, so shouldn't the Moon be pulling harder on the surface molecules of water than on the shore above it? How do we explain why the water appears to defy the Moon's gravity?

First, a basic idea: **materials stretch (or compress) when the force at one end is different from the force at another.** If you and your sister pull on different ends of a rubber hose, the rubber hose will not stretch so long as you are both pulling *in* *the same direction with the same force.* However, once one end is pulled with a force different from the other end, the rubber stretches. The reason?

Force causes acceleration which means a change in velocity. If two parts of a material are pulled with unequal forces, they will

> Matter stretches (or compresses) when one portion is pulled with a different force than another.

accelerate with different rates, leading to different velocities. Obviously, if two regions of a material are moving with unequal velocities, it must either stretch or shrink.

Intermolecular bonds in liquids and solids fight this deformation, in the same way that a spring resists your fingers as you pull one end away from the other. The spring *transfers* some of the force from one end to the other (since it is pulling to the left at one end and to the right at the other). At some point the spring pulls your fingers inward with a force equal to the difference in force at its ends, and the spring stops stretching. When a solid or liquid is being unevenly stretched, it acts like a very, very stiff spring. Small changes in length cancel large differences in force so large differences in force cause only small changes in volume. For this reason, liquids and solids are considered *incompressible*. Liquids are less incompressible than solids. They will stretch a bit more before canceling the difference in force. Because the ocean is so large, this "bit more" that liquids stretch is sizeable enough for us to observe.

> Liquids typically must stretch more than solids to compensate for unequal external forces.

A typical depth for the ocean is 2.5 miles (4 kilometers). Imagine two layers on the far side of Earth: one on the surface, the other 2.5 miles below it. The Moon's gravity does not pull on them equally. The deeper layer is closer to the Moon, so it feels

Effects of Structural Forces on Shore Opposite Moon
(After Gravitational Stretching)

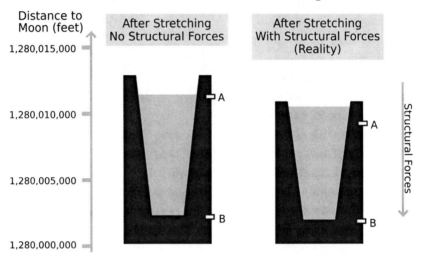

The figure on the left shows what the far shore might look like if matter could freely stretch. However, the structural forces in solids and liquids work to minimize changes in their volume. Here they resist the stretching by pulling the upper material down so that less deformation occurs. Since solids resist deformation more than liquids, the shore is pulled down more than the water at the surface, making it appear to those on the land that the water has risen. Note: this figure is not to scale and is exaggerated for illustration.

a stronger pull. This difference[2] in forces stretches the matter between the two layers, but the water stretches more than the rock. Hence, its surface rises *relative to the shore.*

Low tides: Enter the Pythagorean theorem
The above description explains why a high tide exists, but says nothing about low tide. You might get the impression that

---

[2] It is the *difference in forces* that causes matter to stretch (or compress), *not the force itself.*

low tides are just what happens wherever the Moon's gravity is not a factor, but there is something more interesting going on owing to the Pythagorean theorem.

The strength of the gravitational force between any two objects depends on the square of the distance between them. In the case of Earth and the Moon, this means the gravitational force felt on the Moon-side of Earth is about one fifth of one percent of one percent (0.002%) greater than the gravity felt at the bottom of the ocean at a typical point.

But what happens at a low-tide region, a point halfway between the shore near the Moon and the shore far away? There, the ocean depth is perpendicular (sideways) to the line from the middle of Earth to the Moon, and that is when a consequence of the Pythagorean theorem becomes important.

To see how the Pythagorean theorem is relevant, imagine your sister wants you to take a picture of her in the front yard late one afternoon in June. She hands you her camera and walks 144 steps to the east, figuring it would be best to be facing the Sun (which is in the west). When she stops, she notices that the Sun is not directly behind you. She has to walk 17 steps north for her to see the Sun just over your head. She was obviously 144 steps away from you before she changed direction. How much actual distance did the 17 northward steps add?

As you may recall, the Pythagorean theorem states that for a right triangle, $A^2 + B^2 = C^2$, where A and B are the *legs* and C is the *hypotenuse*. If A is the east/west distance between you two and B is the north/south, then $C = \sqrt{A^2 + B^2}$ will be the actual distance.

The interesting point about you and your sister is that she is mostly east of you, so A (144) is much larger than B (17). Then A^2 (144 x 144 = 20736) is going to be *huge* compared with B^2 (17 x 17 = 289). Since $C = \sqrt{A^2 + B^2}$, and B^2 is tiny relative to A^2, we find that C (the actual distance) is approximately just equal to A (the east/west distance). The size of B has almost

Pythagorean Theorem: $A^2 + B^2 = C^2$

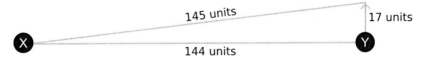

145 units

17 units

X

144 units

Y

If two objects are far apart horizontally, even significant changes in their vertical positions may only mildly affect the distance between them. Point Y could move 17 units vertically without changing its distance from point X by more than 1 unit.

no effect. Indeed, in this case $C = \sqrt{20736 + 289} = 145$, only 1 unit longer than A. So the 17 units of northward movement made very little difference in the total distance because it was "sideways" to the line connecting the two of you.

The point I am trying to make with this example is this:

Due to a consequence of the Pythagorean theorem, the distance between two objects is not substantially changed when either is moved sideways relative to their separation.[3]

To apply this observation to the Moon and tides, recall that the top of the ocean, the bottom of the ocean, and the Moon are all roughly *aligned* at a region of high tide, but they are not aligned at a region of low tide. At low tide the line going from the top of the ocean to the bottom is roughly perpendicular to the line going from Earth to the Moon. This means the change in the distance to the Moon as you go from the bottom of the ocean to the top is significantly less in a region of low tide compared with a region of high tide. Instead of being 0.002%

---

[3]Of course, this statement has its limits. In particular, the direction that was "sideways" when an object began moving will become less and less "sideways" as the object moves. In the diagram, for example, point Y's upward movement is less and less perpendicular to the line between it and X as it travels north.

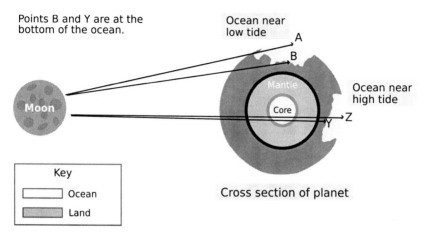

Points B and Y are at the bottom of the ocean.

Ocean near low tide

Ocean near high tide

Moon

Mantle

Core

A

B

Z

Y

Key

Ocean

Land

Cross section of planet

Near points at low tide, the bottom of the ocean and the top of the ocean are almost exactly the same distance from the Moon, so the **strength** *of the Moon's gravity varies little between these spots. However the* **direction** *of that force differs significantly more. The opposite is true at regions of high tide. There, the direction of the forces are practically the same but the distance, and hence the strength of the force, differs appreciably between the top of the ocean and the bottom.*

greater at the bottom of the ocean, the difference in force is only 0.00000001%. In these regions the *strength* of gravity does not vary appreciably even between the top of the ocean and the bottom, but the *direction* does.

Consider again an area on the far side of Earth. Compare a fish swimming at the surface with a giant worm living at the bottom. The *direction* of the gravitational force felt by these creatures is the same, but the strength is different. A 0.002% difference in strength was enough to cause the high tide we have already discussed. If we consider instead the same two creatures in an "in between" (low-tide) region, we see the opposite. There is practically no difference in strength of the Moon's gravity on each, but there is a difference of 0.001% in the *direction* of

The Pythagorean Theorem's Application to Tides

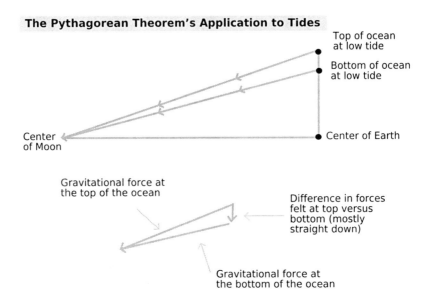

This figure is also not to scale, but illustrates how water in low-tide regions is pressed downward by the difference in the direction of gravitational forces. The strengths of the forces (the length of the arrows) are nearly the same, but the water is pressed inward because the gravitational force pulling on the top cuts a steeper angle than the gravitational force pulling on the bottom.

gravity. The water at the top of the ocean is pushed a little more inward than the water at the bottom.

Since the water at the top of the ocean is pushed more inward than the water at the bottom, the water compresses slightly. The land is compressed as well, but rock does not compress under pressure as much as water does.

So where the Moon is overhead, variations in its gravitational force pull outward, and they push inward where it is near a 90-degree angle. These forces combine with structural forces—the land resists more than the water—to create tides on any planet whose surface has both liquid and solid elements.

82

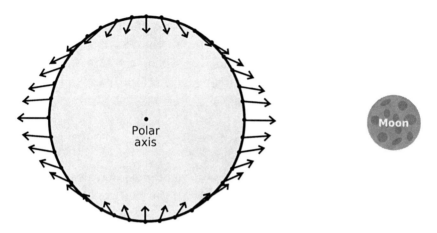

*The figure shows the compression and stretching due to gravity. The arrows describe the **difference** in gravitational force between a point at the top of the ocean and a point at the bottom. At points of high tide, this creates outward stretching because points near the Moon are pulled toward it more strongly than points away from the Moon. Near places of low tide, compression occurs because the force of gravity at the top of the ocean cuts more sharply inward than the force of gravity at the bottom. This image was inspired by a similar one on Bill Otto's Bad Science webpage.*

But on Earth, tides are much higher than what one would expect based solely on these factors. Our tides are amplified by the ability of water to flow from one point to another.

Enter fluidity

The surface of Earth is largely covered with water, and almost all of it is connected. This amplifies the effect described in the last section.

In the last section we noted how water in a low-tide area is pushed inward. If these low-tide areas were simply vast lakes, that would be the end of it. However, water in Earth's oceans is not captive to its local basin. At some point it becomes easier for the water to flow away rather than be compressed against deeper water. Thus, water in the low-tide areas will slosh toward the high-tide areas, increasing the total amount of water in those regions. This draining of water from one region to another greatly increases the size of tides on Earth.

Conclusion

Because water can stretch (and compress) more easily than land, any difference in force between the bottom of the ocean and the top can cause tides. The Moon is the principal cause of tides on Earth. High tides are generated where differences in *distance* cause a considerable difference in the *strength* of gravity between the top and bottom of the oceans. Regions of low tide occur where differences in the *direction* of gravitational force presses the water toward the center of Earth.

While gravity is the fundamental *source* of tides on Earth, the height of tides is heavily enhanced by the interconnectivity of its oceans. Water in a low-tide region can drain into high-tide regions, greatly increasing the size of the tides.

Tides are not in any way dependent on or due to the orbital motion of Earth in the Earth-Moon system. We would experience tides even if this motion were stopped and the Moon dropped straight down toward Earth. We just would not get to experience them for long.

Before diving into a discussion about the widely misunderstood and misnamed "greenhouse effect," I want to clarify that this chapter does not focus on global warming or the effect humans have on climate. I will discuss the controversy a bit, but my chief interest lies in the explanation given by textbooks, not the larger questions concerning the reality and consequences of the global warming. No educated scientist (whether skeptical of anthropogenic global warming or not) disagrees with the general principle that our atmosphere increases Earth's temperature.

The standard explanation goes something like this:

> Atmospheric gases act like the walls of a greenhouse, trapping most of the energy given off by Earth's surface and (re)radiating a portion back down. The net result is that Earth's surface receives more energy in total than it would were it lacking an atmosphere. This is how an atmosphere increases the average surface temperature of a planet.

Sometimes this explanation is further simplified to say the atmosphere traps heat waves coming from Earth's surface. Such explanations are deficient because it is Earth's surface temperature that most people care about, and the atmosphere's absorbing heat would have no relevance unless that heat got sent back to Earth somehow.

You may have already guessed from my use of quotation marks that one problem with this explanation lies in the misconceived analogy to a greenhouse. Unfortunately, a hundred years of scientists' trying to disabuse the population of this imagery has hitherto proved unsuccessful. The comparison is truly ironic; the mechanism by which our atmosphere influences Earth's average temperature is, in some sense, the exact opposite of how warmth is generated in a greenhouse. Indeed, the *Moon's* surface becomes unbearably hot owing to an authentic greenhouse effect precisely because it *lacks* an atmosphere.

Most everyday examples of heat exchange, including the mechanics of a greenhouse, are examples of *convection* and/or *conduction*. The heating and cooling of a planet is almost entirely driven by a different mechanism: *radiation*.

It has been vocally argued in education circles that the greenhouse analogy is misleading and inaccurate, but less widely realized is that there is a second reason why atmospheres lead to higher average surface temperatures for the planets they envelope. This second cause has little to do with absorption of infrared radiation.

These two issues are largely independent from one another, and each deserves its own investigation, so the balance of this chapter is divided into two sections. The first is longer, reviewing the basic methods of heat transfer before investigating the "greenhouse effect" in detail. The shorter second section discusses the *other* way an atmosphere increases the average temperature of a planet's surface. In Earth's case this second, little-trumpeted mechanism increases temperatures more than the "greenhouse effect" that gets so much attention.

An Outdated Metaphor

A quick overview of the three standard *modes of heat transfer* may be useful.

Most of the heat transfer we experience inside our homes involves *conduction* (the transmission of heat through a substance) or *convection* (the transfer of thermal energy by moving matter). When you warm milk in a pot on the stove, the air above the milk picks up heat and rises to be replaced with colder air; the milk is losing heat by convection. The milk would warm more quickly in a closed pot. Coffee will stay hotter in a thermos, even an open one, than in a metal cup because the metal will conduct heat away from the coffee to be lost to the air. If you close the thermos, you hin-

> Every object radiates heat based on its temperature. We see some of the Sun's radiation as light.

der both convection and conduction, and the coffee can stay hot for a long time. Convection, when possible, is generally a greater source of heat transfer than conduction.

Conduction and convection convey heat by the interaction of atoms in matter. They cannot transmit energy easily through space. Where matter is sparse, heat moves primarily through *radiation* instead. You can think of radiation as "heat waves" given off by objects. *Every object in the universe radiates energy.*

The type of radiation an object naturally emits depends on its temperature. The Sun is hot enough for us to see its radiation as visible light. We do not, however, see the radiation given off by an ice cube, but only because our eyes are not adequately sensitive. If we were built differently (able to see light of lower frequencies), we could see everything around us glowing. The makers of Unisom™ would be rich. I will use the term *light* to refer to this radiation throughout the chapter. Just keep in mind that much of this "light" is not visible to humans.

Scientists used to think greenhouses stayed hot because the glass absorbed radiation given off by the objects inside. In the mid-19th century, when the absorption of heat by our atmosphere was first openly considered by scientists, it matched the way they thought greenhouses worked.

Later, scientists realized they were wrong about what keeps a greenhouse warm, but the name "greenhouse effect" had already come into being. Rather than discontinue using the metaphor, textbooks simply gloss over the error with semantic sleight-of-hand, convincing students that trapping *energy* (what they say the atmosphere does) is somehow similar to trapping air (what a greenhouse does).

A misunderstanding about how greenhouses function created the term "greenhouse effect." By the time scientists realized they had been wrong about how greenhouses worked, the term was already entrenched.

Therein lies a principal problem with how we teach this topic. Suggesting that "trapping air" is in any way similar to "trapping heat" causes mental whiplash for students. A pupil first learning about the three standard modes of heat transfer (conduction, convection, and radiation) is taught to see these two scenarios as entirely different because they deal with two disparate kinds of transfer. "Trapping air" refers to *preventing convection* while "trapping heat" refers to *enhancing radiation*. Middle-school teachers ask students to cognitively separate these mechanisms, and anything that conflates them imperils this goal. A further problem is that "greenhouse" gases do not trap energy at all. They merely slow how quickly the energy leaves.

Later in this section I will give a more apt analogy that does not confuse one essential idea with another. First though, let's look closely at each mechanism.

Warming of celestial objects

Large bodies in space, like planets and moons, receive energy in the form of radiation. Their surfaces also emit radiation based on their temperature. *A spot on the surface will heat up or cool down until the heat it is losing (through radiation, convection, evaporation, etc.) equals the heat it is receiving.*

Let's apply this to the Moon. The Sun emits radiation, some of which hits the Moon. The Moon absorbs most of this light, reflecting a bit to ornament our night sky. Since the Moon has no atmosphere, there is no wind to move heat to a cooler region and no water to cool the surface by evaporation. So a surface region in direct sunlight will rise in temperature until it is hot enough to radiate away as much energy as it receives. The Sun-side of the Moon heats to a blistering 225° F (107° C) before it sends out heat as rapidly as it absorbs it.

Thus it is the Moon, rather than Earth, that endures a true greenhouse effect, precisely because it *lacks* an atmosphere. Objects in a greenhouse enjoy no cooling by outside air currents, and humidity inside opposes evaporation. Areas on the lunar surface in direct sunlight suffer high temperatures for the same reasons. Absent an atmosphere, there can be no water to evaporate and no wind to move heat.

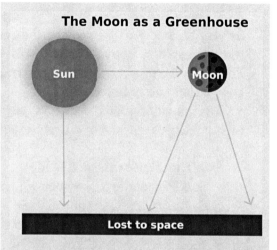

The Moon as a Greenhouse

Sun → Moon

Lost to space

There is a true greenhouse effect on the Moon specifically because it lacks an atmosphere. There is no wind on the Moon to blow the heat from hotter regions to colder ones (as wind does on Earth) and no water to cool the surface through evaporation. The space inside a greenhouse, which also has no direct contact with the outside air, warms for the same reason.

Our atmosphere[1] complicates the situation on Earth. The gases, and in particular the clouds they form, reflect a significant

---

[1] I will often use "atmosphere" to refer to "atmosphere plus clouds." Climatologists often separate these, but clouds would not exist on Earth without our atmosphere, so for my purposes I will combine them when discussing the effect an atmosphere has on a planet.

percentage of the sunlight coming our way. They also absorb some of the light radiated by the Sun and Earth. Most importantly for our discussion, the atmosphere (just like everything else) sends out its own "light." A portion of this energy makes its way back to Earth's surface; the rest is lost to space. Finally, the atmosphere allows other forms of heat transfer to leach heat from Earth's surface, like evaporation of surface water and convection due to breezes, etc.

We can categorize the effects described above into one that warms Earth's surface and three others that cool it:

One effect that warms Earth:
1. The atmosphere provides a separate energy source for Earth. It, like all other matter, sends out radiation, some of which is directed downward to the surface.

Three effects that cool Earth:
1. The atmosphere reflects some of the sunlight that would otherwise have hit Earth.

2. The atmosphere absorbs some of the sunlight that would otherwise have hit Earth.

3. The atmosphere cools Earth by allowing convection and evaporation.

While saying *atmospheric gases heat Earth's surface by absorbing heat* is much better than referring to a greenhouse, these lists show why such a facile explanation is suspect. It is true that the gases absorb radiation, but they only re-radiate a portion back down to Earth. The energy is not trapped in an endless loop. The common explanation of the greenhouse effect does not make at all clear why the amount of extra radiation coming from the atmosphere outstrips the combined strength of the three cooling effects.

The Heating and Cooling of Earth

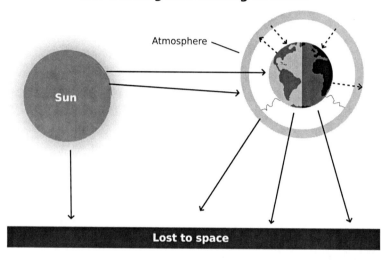

Atmosphere

Sun

Lost to space

- 〜〜 Non-radiative cooling (e.g.,convection, evaporation)
- ‑‑‑> Radiation related to "greenhouse" effect
- ——> Other Radiation

Earth's atmosphere is shown as a ring but actually reaches all the way to Earth's surface. It receives energy from both the Sun and Earth, and it also radiates energy away both toward Earth and out to space. The atmosphere provides a second source of heat for Earth's surface but also helps hot regions on Earth's surface offload heat through wind and evaporation. Thus, an atmosphere cools the day-side of Earth much like opening a greenhouse door would cool the inside. The atmosphere prevents the true greenhouse effect we see on the Moon by allowing redistribution of heat.

Whether the atmosphere of a planet increases or decreases the energy received by the surface depends on properties of its star and the composition of the atmosphere. Two facts conspire to enhance the heating mechanisms in Earth's case. First, the Sun's temperature is such that a significant portion of its radiation penetrates the atmosphere without being reflected or absorbed. Second, the atmosphere is layered in such a way that it sends more energy downward than upward.

> Discussions of the "greenhouse effect" tend to ignore the cooling aspects of an atmosphere, and so forgo a rich inquiry into why these do not compensate for the extra heat received.

Different gases absorb different wavelengths ("colors") of light depending on their molecular properties. Water vapor is the only greenhouse gas that absorbs much visible light, and its absorption is limited. If the atmosphere absorbed significant visible light, the Sun would look dim.[2] About half of the solar energy aimed at Earth makes it to the surface simply because the gases are not equipped to absorb or reflect it. The Earth's surface (land and water) absorbs about 96% of that light.

Air particles nearest the Earth are warmer (and hence radiate more) than average. Because radiation is likely to be re-absorbed if it has to travel past lots of molecules, the consequence of having warmer molecules closer to the surface is that it is easier for downward radiation to make it to the surface than it is for upward radiation to escape. If our atmosphere had a different composition, its temperature profile would also be different, and we might not receive quite as high a proportion of the energy radiated by the atmosphere. Titan, Saturn's largest moon, experiences a weakened warming effect because of this.

---

[2] Clouds block the Sun, but not by absorbing the light. They *scatter* it. Very tall clouds look dark at the bottom as less and less light penetrates directly.

Heating of a greenhouse

For over a century, scientists have known that the air inside a greenhouse is warmer because the glass structure stops heat loss by convection, not because it enhances heat gain by radiation. The fact that the glass absorbs the mild infrared radiation coming from the objects inside is immaterial.

There are two easy ways to show that the absorption of radiation by the glass of a greenhouse is not a significant factor:

- The effect persists even when the greenhouse is constructed from material that does not absorb infrared light, like specially manufactured glass or rock salt.

- If a small window is opened, the temperature sharply decreases. Were greenhouses warmed by glass (re)radiating infrared light, opening a window would have negligible effect. Conversely, a "hole" in the atmosphere, a region where the gases did not absorb light, would not cause the kind of rapid cooling observed when a window is opened in a greenhouse.

The black shirt effect?

Unfortunately, scientists discovered the atmospheric effect prior to realizing it differed from the mechanism that warmed greenhouses, so the misconceptions of 19th-century scientists are to blame for our 21st-century misnomers. Everyday life provides more appropriate analogies to each of these situations, and these may clarify the differences between them.

Our body naturally heats nearby air. If a breeze blows this air away, it cools the skin because new air replacing it is cooler and can leach more energy away. Contrariwise, a blanket keeps us warm primarily by trapping this warmed air. This is why even a thin sheet can be effective for temperature regulation. Thus, warming plants with a greenhouse is similar to the use of a blanket for warmth.

Rather than retaining warm air, as blankets and greenhouses do, our atmosphere causes Earth to be warmer than otherwise for much the same reason that a dark shirt keeps you warmer than a white one. In addition to absorbing most visible light, typical black dyes also absorb infrared rays emitted by our bodies. Thus, like the atmosphere, a black shirt provides a second heat source for your body, being warmed by radiation from both sides.[3]

> A real greenhouse keeps air warm similar to the way a blanket does. What we call the "greenhouse effect" is more like wearing a black shirt instead of a white one.

Warming by Moderation

I indicated at the beginning of the chapter that there are two basic reasons why an atmosphere causes a planet to have higher average temperatures. The last several pages described one of them. Now, let's look at the second reason, which no one mentions even though it is responsible for more warming of Earth than the "greenhouse effect."[4]

Most of the warming experienced by Earth's surface is the logical consequence of connecting the following two statements:

---

[3] Manufacturers of sports apparel produce fabrics with different infrared absorbances depending on the needs of players. Purveyors of alternative healing techniques have even gotten into the gig, marketing clothing alleged to possess healing properties due to absorption and emission of infrared rays.

[4] It is hard to put down a precise number describing how much is attributable to each, mostly because we have such a limited understanding of clouds. People often claim that turning off the "greenhouse" effect would cool down the earth by 60° F (33° C), but such numbers are dubious because they assume "turning off" the greenhouse effect would have no effect on clouds, which is ludicrous. It's fair to say the greenhouse effect increases the temperature on Earth by at least 18° F (10° C) and probably significantly more. The mechanism investigated in this section increases average temperatures by 70–90° F (40–50° C).

1. An atmosphere *moderates* temperatures around the globe.

2. Moderating the temperature of a planet causes its average temperature to increase.

1. Atmospheres moderate surface temperatures

The daytime temperature on Mercury's equator is about 800° F (425° C) while the temperature in the same place at night is -280° F (-170° C). On the other hand, Venus' temperature does not vary much at all. From pole to equator, day to night, Venus' surface is about the same temperature everywhere.

Why does the planet nearest the Sun suffer the greatest temperature extremes while the second one has a temperature distribution even more uniform than Earth's? It's simple. Mercury has practically no atmosphere while Venus' atmosphere is nearly 100 times more massive than Earth's. The maximum temperature difference between any two points on Earth is about 235° F (130° C). Compare that with the Moon, which is approximately the same distance from the Sun but where the maximum temperature difference between any two points is about 650° F (360° C).

An atmosphere serves to moderate temperatures in several ways, many of them having nothing to do with absorption of radiation. For example, wind allows hot regions to warm colder ones. Our oceans would not exist without an atmosphere.[5] Oceans do a fantastic job of moderating temperatures both by storing thermal energy and moving that energy around.

---

[5]Liquid water requires an atmosphere for the same reason that a pressure cooker speeds up food preparation. The more pressure a liquid is under, the higher its boiling point. The water in a pressure cooker reaches a hotter temperature before boiling than the water in an open pot, so food cooks faster in it. Similarly, the less pressure water is under, the lower its boiling point. On Mars, water would boil immediately because its atmosphere is so thin. There is also the separate point that the absence of an atmosphere precludes condensation, so net evaporation rates would be sky high.

2. Moderation raises average temperature

The moderating effect of an atmosphere is nothing controversial, but you may find it counter-intuitive that spreading thermal energy around actually increases a planet's average temperature. When we speak of "moderating" temperature, we generally think of the temperature rising in one place as it decreases in another.

> A planet without an atmosphere cannot have liquid water on its surface.

Recall that planets heat up until they reach *thermal equilibrium*. At that point the amount of energy they give off is equal to the energy they receive. Hot surfaces give off *much* more heat than cold surfaces. It is not a linear relation. If one surface is twice as hot as another, it actually radiates *sixteen* times as much energy per second!

What does it mean for one surface to be "twice as hot" as another? Scientists use the Kelvin scale to talk about absolute temperature. The lowest possible temperature is *absolute zero*, or 0 K. A kelvin[6] represents the same change in temperature as a degree Celsius. Room temperature is about 300 K. When I refer to something being "twice as hot" I mean one has a temperature of twice the other on the Kelvin scale.

Let's say a surface whose temperature is 100 K gives off 1 gagglehooch of heat per square meter. I have made up this unit, the *gagglehooch*, to simplify our discussion. To convert to standard units, 1 gagglehooch = 1.38 calories per second. Then the following table shows how many gagglehoochem of heat surfaces at various temperatures radiate per square meter. The relationship is $H = (T/100)^4$.

---

[6]Similar to "newton," the *unit* of temperature called "the kelvin" is not capitalized. Nor should one speak of "degrees" Kelvin. The unit of the Kelvin scale is the kelvin.

Temperature	Heat Radiated (per Second)
100K	1 per square meter
200K	16 per square meter
300K	81 per square meter
400K	256 per square meter
500K	625 per square meter
600K	1296 per square meter

Imagine a ball with a surface area of 2 square meters. If the ball's surface is 500 K, it gives off 625×2 = 1,250 gaggle-hoochem of energy per second. This ball models a planet with moderated temperatures—we pretend the ball is the same temperature everywhere.

Compare that with a ball where half is very hot (600 K) and the other half is very cold (100 K). This ball gives off 1,296 gagglehoochem of energy from its hot side and 1 from its cold side, giving a total release of 1,297 gagglehoochem per second. Thus, it gives off about the same total energy as the first ball, but the *average* surface temperature is much less (350K instead of 500K).

The hot side of the second ball is only 20% hotter than the first ball, but it radiates more than twice as much energy per square meter. This more than makes up for its cold side, which radiates almost no energy.

The same thing happens with planets. Planets that have an atmosphere balancing their temperatures have to get hotter on average because they do not radiate energy as efficiently as a planet that has extreme temperatures, whose hotter regions can more than make up for their colder ones. For example, the hottest point on the Moon at any given time is nearly 400 K (260° F / 127° C) while points near its poles, which never receive direct sunlight, can get as cold as 26 K (-413° F / -247° C)! The hot spot is fifteen times hotter than either pole, but it gives off around $15^4 \approx 50,000$ times more energy per square meter. The Moon's scorching day-side radiates such a tremendous amount of heat that the surface as a whole only needs to average $220\,\mathrm{K}$ to offload all the energy it receives. If the surface of the Moon had more moderated temperatures, it would have to be $50\,\mathrm{K}$ hotter on average to radiate away the same amount of energy.

Conclusion

To stay at a steady temperature, a planet must lose energy equal to the amount of radiation it receives. An atmosphere allows it to lose some heat through convection, evaporation, and similar matter-based heat processes. More importantly, it facilitates the transfer of heat from hotter portions of the planet to colder ones. Thus, it would be more accurate to use the label "greenhouse effect" to describe the plight of planets and satellites that *lack* an atmosphere because regions in direct sunlight, like plants in a greenhouse, are not cooled by convection or evaporation. This is why the Moon's surface can get incredibly hot—hot enough to boil water.

There are two principal ways an atmosphere causes the average temperature of a planet to increase, and neither of these reflects the workings of a greenhouse.

First, an atmosphere provides a second source of radiation for the planet. Depending on the composition of the atmosphere and the temperature of the star the planet orbits, this radiation can increase the net radiation received by the planet. Second, an atmosphere serves to moderate temperatures on a planet due to wind and, in the case of Earth, water. Planets with moderated surface temperatures are much less efficient radiators of energy. To radiate away the same amount of energy, their average temperature must become significantly higher than one with extreme temperatures.

Appendix: Global Warming Controversy

As mentioned earlier, my interest is not to go deeply into the miry waters of the current debate over human-induced global warming. However, people are exposed to so many competing voices that I assume most readers would find the omission of the topic strange in a book meant to address commonly misrepresented topics in science.

Upon reading the chapter, one might wonder why there is any debate at all. If gases in the atmosphere radiate energy downward, warming Earth's surface, why is there so much controversy over human-induced global warming? Why have thousands of scientists with PhDs claimed a skeptical stance against thousands of others who have concluded the world is dangerously warming due to human activity?

The simple answer to this question is that carbon dioxide, the gas at the heart of the discussion, can only absorb a small percentage of the various "colors" of light emitted by Earth, and most of those would be absorbed by water vapor even if there were no carbon dioxide in the atmosphere.

By way of analogy, imagine a vast banquet prepared for a wedding. Due to poor planning, the caterer did not realize that all the guests were strict vegans, refusing to eat anything involving the use of animals (meat, eggs, milk, honey, etc.) While there is abundant food, only a small percentage can be eaten by the guests.

If only a few people come, each would likely find plenty of food, and the amount of food at the banquet would decrease steadily. However, as more and more guests arrived, there would be less and less food that met the guests' diets. Eventually, increasing the number of guests would not significantly change the amount of food eaten. Those coming later would find that all the food allowed by their diet had already been consumed.

Similarly, while more carbon dioxide will lead to more absorption of certain "colors" of light, and this will increase the amount of radiation hitting Earth, the effect quickly diminishes as the amount of carbon dioxide increases. There is a limit to how much energy the CO_2 can absorb, regardless of how much CO_2 is in the atmosphere. Since most of the light carbon dioxide can absorb is going to be absorbed anyway by water vapor in our atmosphere, the maximum *direct* warming due to carbon dioxide is very small. Indeed, our atmosphere may already be at the point where additional CO_2 has almost no effect. We don't know because modeling heat exchange among layers within the atmosphere is difficult.

The real question is whether this small amount of direct warming can lead to more indirect heating. In some ways, it certainly can. Ice reflects a good deal of light, keeping Earth cooler. A small amount of direct warming could lead to more as ice melts, decreasing the amount of light reflected away. On the other hand, more warming creates more clouds, which cool Earth. The interconnections are staggering in complexity, and some models suggest Earth regulates itself while others suggest warming begets warming.

It doesn't help that we lack unambiguous temperature data. In particular, scientists always adjust data they consider artificially off (for example, an incinerator may be built near a thermometer that had originally been in an open field), and this process of "homogenizing" the data allows two scientists to reach wildly different conclusions simply by choosing different adjustments. In particular, many scientists believe the observed rise in temperatures comes not from Earth's surface as a whole getting hotter but rather from the simple fact that cities produce a great deal of extra heat. Thermometers near growing cities are expected to record higher temperatures as the city increases in size.

It is known that the temperature of Earth's surface (or at least the Northern Hemisphere) over the last millennium has varied between being both much colder (causing a "little ice age" a few centuries back) and much hotter than today (allowing Greenland to be ... well ... green, supporting its Norse inhabitants during the Vikings' heyday). Indeed, environmental scientists of all stripes agree that Earth's surface is currently colder than average when one looks back millions of years. They see our present age as the end of an ice age; for much of Earth's past there was no ice even at the poles.

However, the existence of warmer periods in the past does not completely invalidate modern concerns. If the *causes* for those previous warmer periods are different from the current one, then the aftermath may be as well. Furthermore, just because Earth's natural average temperature (over geologic time) is hotter than today's does not negate the obvious problems a warming Earth has for human civilization.

The Coriolis effect, which refers to the tendency of a moving object's path to *appear* to curve due to Earth's rotation, is often triply slighted in the science classroom. First, it is frequently incorrectly cited as the reason water draining from a sink in the Northern Hemisphere swirls the opposite direction as it does in the Southern Hemisphere. Secondly, students receive mixed messages regarding phenomena the Coriolis effect does actually influence. For example, students are told Coriolis causes flying objects to turn to the right[1] in the Northern Hemisphere, but they are also told Coriolis causes hurricanes, which spin counterclockwise. Finally, and most significantly, explanations for how the effect arises can be cryptic.

I provide two explanations for the Coriolis effect in this chapter. An addendum addresses the mixed messages students receive about how Coriolis, a "fake force," affects the natural world. To make short work of the first insult: no, the direction water twists as it is drains does not depend on which hemisphere you are in. The Coriolis effect is far too weak to dictate the direction of swirl in any commercial toilet or bathtub. The behavior of draining water in such household contexts depends primarily on the shape of the basin. On the other hand, Coriolis does determine the direction hurricanes rotate, the general pattern of ocean currents, and the prevailing winds on a global scale. Therefore, it should be considered important notwithstanding its inability to visibly affect your bath water.

Before describing what causes Coriolis, let's take a look at some common, unsatisfactory explanations.

---

[1] I will be using phrases like "curve to the right" several times in this chapter. Unless clearly stated otherwise, I mean that someone at the location of the object, watching it fly away from behind, sees it curve in that manner. In other words, it curves to *its* right.

"Earth Rotates Under Us"

The worst explanation claims the Coriolis effect is due to Earth's rotating under the object while it is in flight. Students given this explanation are led to believe that flying items appear to curve because we observers are moving with Earth, and the flying object is not. Such a narrative is wonderfully simple and would make this whole topic extremely easy to understand, but is also very wrong.

It is not hard to see why this explanation cannot be correct. By its logic, all items should curve to the west because Earth (and all observers fixed on it) rotates eastward. But the Coriolis effect can cause items to curve in any direction. Indeed, hurricanes could not be attributed to Coriolis if the curvature were always westward. Air particles in a hurricane are forced to go in rapid spirals, and this circular motion cannot be accomplished by westward deflection alone. Furthermore, this explanation would indicate the Coriolis effect is greatest at the equator, where the rate of revolution is highest. In reality, there is no Coriolis effect at the equator.

Let's look at a concrete example. Consider a quarterback in Dallas, Texas, throwing a football north. The Coriolis effect applies to *any* moving object, including flying footballs. If Coriolis were due to our eastern rotation in the way described by this explanation, the football would curve westward at 400 meters per second. Receivers would be the highest paid players in football and might even deserve their salary, assuming any human could reliably catch such drastically curving pigskins.

To see where this explanation goes wrong, consider the instant the football is thrown. The quarterback, the fans, the

> No Coriolis exists at the equator, yet that is where Earth's rotation is fastest.

referees, and everyone else is moving eastward with the Earth at around 400 meters per second. We do not feel this motion for

the same reason we can drive very fast on an interstate without severe discomfort. If everything is moving at the same velocity, including the air particles surrounding us, there is nothing to indicate we are moving.

The quarterback, receiver, fans, and ref are not the only items moving eastward at 400 meters per second. The ball in the quarterback's hand also has that velocity. When he lets go of it, the ball is moving at the same speed as Earth's surface, and it continues at that speed while flying. This is why he can throw the football straight: it is moving just as fast as Earth's surface.

The same applies to water in the oceans and air in the atmosphere. Just because a molecule is not touching land does not mean it is untethered to Earth, unaffected by its rotation. Air near Earth's surface rotates at about the same speed as the land beneath. Otherwise, we would constantly feel a devastating wind as Earth's rotation rammed us into stationary air.[2]

---

[2]The force of gravity coupled with *viscosity* causes air near Earth (and water in its oceans) to move with the rest of us. Viscosity is discussed at length in the *Flight & Bernoulli's Principle* chapter, found in volume 2.

Furthermore, it is possible to have rotation with no Coriolis effect. If Earth were a spinning cylinder instead of a spinning sphere, there would be no Coriolis effect, at least not for objects moving horizontally.[3] The key difference between a spinning cylinder and a sphere is that an object on a spinning cylinder always stays the same distance from the axis of rotation. *If you only consider horizontal motion, the Coriolis effect is always due to an object getting closer to or farther from Earth's axis.*

Why does the distance from the axis matter? That is the subject of the second common explanation.

"Rotating at Different Rates"

The first explanation failed because objects near Earth's surface typically rotate with Earth, even if they are not fixed to it. When I throw a football, Earth rotates under it with approximately the same speed as it did when it left my hand.

The second common explanation picks at the "approximately" in that last sentence. It attributes the Coriolis effect to the difference in rotational speed between one location and another. Earth's rotation moves people at the equator faster to the east than those at any other latitude because they have farther to travel each day. An object thrown due south from the equator begins its journey with that eastward velocity. If it travels north or south, it finds itself going faster eastward than the terrain below it. Surface-bound observers see it bank eastward, so it curves to the right if flying north and to the left if flying south.

The above explanation is much better than the first. Unfortunately, it still has significant problems. Most worrisome is the suggestion that the Coriolis effect has no bearing on an object moving east or west. It is clear why a ball moving north or

---

[3]Coriolis can cause objects to appear to curve upward or downward, but gravity has such a great effect on the height of a flying object that we normally discount any Coriolis effect in these directions.

south will pass over terrain moving eastward at different speeds, but points along the same line of latitude rotate at the same velocity. For a bird flying along the Tropic of Capricorn, the terrain beneath it is moving at the same speed when it begins its journey as when it ends.

> Many explanations for the Coriolis effect fail to adequately explain why objects moving due east or west also appear to curve.

This implication that movement due east/west is not affected by Coriolis is a severe drawback. Most natural phenomena materially influenced by the Coriolis effect involve particles making circular journeys (e.g., air molecules in a hurricane, water particles moving in vast, circular currents, or global wind cycles). To accomplish these circular journeys, Coriolis has to be able to act on a particle at all times, which means it must be effective for all directions of travel. If Coriolis only affected particles moving north or south, hurricanes could not form.

There is a further difficulty with this explanation. While it is true that the Coriolis effect can be attributed to the fact that the eastern velocity of the underlying terrain changes as an object moves north or south, that tidy explanation masks some real subtleties concerning motion on Earth's surface. This would not be a problem in itself; sometimes a solution that sidesteps subtleties is superior because it is easier to understand even if it is not quite as rich. In this case, however, the nuances masked by the succinct explanation are exactly the aspects that help us understand why the Coriolis effect works for east/west motion as well.

I elaborate later, but here I'll simply say that compensating for the change in velocity from one latitude to the next does not suffice to make an object appear to travel straight. Imagine an arrow is shot from the North Pole. As it travels from the Pole, it passes terrain moving faster and faster eastward. Now pretend this arrow has special rockets that give it a slight push to the left that exactly equals the change in rotational speed. The *Rotating at Different Rates* explanation implies that such an arrow would appear to fly straight because the rockets make up for the changing rotational speed. As we shall see, such a rocket-powered arrow still appears to curve, though only half as much as a mundane one.

"Moving Target"

Another faulty explanation, similar to the last, focuses on the target of a moving object rather than its mid-flight perspective. Consider a hunter at the North Pole[4] shooting at a bear in Greenland, 1200 miles away. The bear (stationary relative to the ground) is moving about 300 miles an hour eastward while the person at the North Pole is not moving eastward at all. The hunter is simply spinning slowly at the pole. While the hunter's arrow flies the 1,200 miles, the bear is moving to the hunter's left, so the arrow misses.

The hunter shot in the direction he was facing, and he continued to face the bear the entire time. Logically, if the arrow misses the bear, we can then deduce that the hunter saw the arrow deflect. Someone viewing from space would describe it differently. She sees the bear move to the left of the hunter's original line of sight.

This explanation suffers the same major drawback as the last. It is pretty convincing at explaining why items curve when moving north or south, but implies no curvature for objects moving east or west.

---

[4] Earth's surface is roughly a sphere, so it slopes downward, away from all observers. For all of the discussion regarding the North Pole, I am ignoring this downward curve. For illustration purposes, someone at the North Pole is like someone at the center of a flat turntable. The mathematical complexities required to account for the downward slope do not alter the explanation.

Additionally, this focus on the target veils the flying object's point of view, which can lead someone unwittingly to error. For example, one could mistakenly believe that the solution for the hunter is to shoot at an angle so his arrow flies 300 miles per hour to the left, matching the velocity of the bear. Leading the bear in this way would nullify the difference in velocity, but the quarry would still be safe. The bear is moving in a large circle. It might be moving 300 miles per hour eastward all the time, but "eastward" changes from point to point. It only matches the arrow's speed at the beginning. Overleaf, I have placed a diagram comparing these two situations for an arrow expected to take 6 hours to reach the bear. The third and fourth figures illustrate the effect of the bear's circular path.

Roadmap

Aside from the conclusion and addendum, the balance of this chapter is segmented into three parts: *Account 1: Everywhere is the North Pole*; *Interlude: The Full Story at the North Pole*; and *Account 2: Curving Latitudes*.

Those who just want a short explanation, without knowing much about what is going on under the hood, should find the first of these sections useful. Those who want a deeper understanding should find their needs met by reading the first two sections. Readers who do not find the first explanation satisfying, or those who would like to know two different explanations, will want to read all three.

Account 1: Everywhere is the North Pole

The *Moving Target* explanation would be useful except it wrongly implies the Coriolis effect only influences objects moving north or south. If I aim at an object due east, it is rotating with the same speed I am.

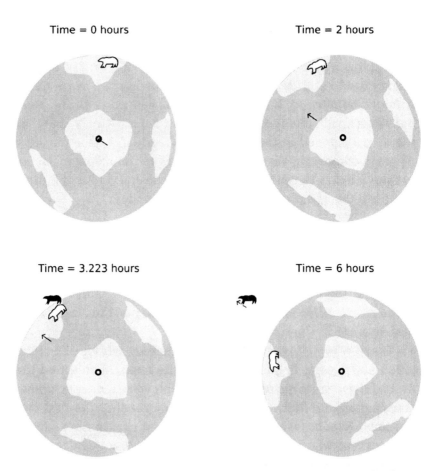

Time = 0 hours Time = 2 hours

Time = 3.223 hours Time = 6 hours

An arrow shot with leftward velocity matching the initial speed of the bear misses the prey. The shadow-bear in the bottom two snapshots shows where the bear would have been if it had kept its initial speed without Earth turning it in a circle.

This difficulty does not arise when describing an observer at the North Pole because there is no east or west there. Everything is south. Any target will be moving to the observer's left. Hence, the *Moving Target* explanation works for all directions for someone at the North Pole. If we can convince ourselves that every spot on Earth is similar to the North Pole, at least in all the ways relevant to the *Moving Target* explanation, we will have arrived at an acceptable account of the Coriolis effect.

Let's consider what makes the North Pole interesting. Two observations present themselves immediately. First, someone standing at the North Pole turns at the same angular rate as Earth, spinning around once per day. Second, other than this spinning, Earth's rotation does not move someone standing at the pole. For the same reason that people on the equator move the fastest, those at the poles move not at all.

Regarding the first point, the spinning imparted to an observer is undoubtedly important. The whole idea behind the Coriolis effect is that rotation causes observers to see things differently. The actual speed of the rotation is not important. Someone at the North Pole rotates once every 24 hours, but the same effect would occur (to greater or lesser extents) with faster or slower spins.

The second point, that the North Pole is not moving, is immaterial! Observers are always assumed to be immobile from their own perspectives. If they moved, it would not destroy the Coriolis effect. So long as Earth was still spinning them around, no other point could appear stationary. If no other point appears stationary, then every other point is a moving target.

Imagine anchoring a steel post precisely at the North Pole. Now cut a hole in the middle of a large disk, and slip the disk onto the post. You have made a type of merry-go-round. If there is no friction between the disk and Earth or the disk and the post, someone standing on this merry-go-round sees Earth spin beneath her. This is how we might envision Earth looking from a spacecraft hovering above the North Pole. Normally, someone standing at the North Pole would spin because of friction between the soles of her shoes and Earth's surface. But someone on a frictionless merry-go-round would be immune to Earth's spin.

Imagining standing on these special merry-go-rounds at other locations help us understand Earth's motion because they allow us to separate the *translational* motion (moving from one location to another) from the *rotational* motion. The merry-go-round's pole is anchored, so it moves eastward with Earth's surface, but its disk cannot be twisted by Earth's rotation.

Not counting the equator, where there is no Coriolis effect, every point on Earth's surface is like the North Pole in that an observer rooted there is spun by Earth. This means anyone standing on a frictionless merry-go-round sees the world rotate beneath him. People near either pole see Earth spin under them nearly once per day, counterclockwise near the North Pole and clockwise near the South Pole. Those living near the equator see Earth spinning less rapidly. Spectators, fixed to the ground, perceive the merry-go-round as moving, but in reality Earth is rotating them around it.

Star trails above Glastonbury Tor, fabled resting place of King Arthur.

The rest of this section is devoted to convincing you that Earth really does spin us in this manner. If we clung to a torsionless rope attached without friction to a tree, we would see everything slowly rotate around us.[5] Those who think they are the center of the universe might have some justification.

Curving stars

The peaceful pastime of stargazing provides an excellent way to observe that Earth spins us. Face due east, pick a single star near the horizon, and firmly set your feet so you are facing it. Unless you are on the equator, you will find the stars curve. Their bending paths cannot be attributed to the simple eastward

---

[5] Here, torsionless means the rope does not resist being twisted. Note that the rope in this thought experiment is *attached to the tree* without friction, but you would certainly have to rely on friction between your body and the rope to cling to it.

rotational motion of Earth beneath you. Since you are facing east, more eastward motion should only cause stars to move straight up. Instead, they rise and curve.

The stars rise because Earth is moving you east. They curve because Earth is also spinning you. Those in the Southern Hemisphere see the universe spinning clockwise about them while those in the Northern see the opposite.

Now imagine you are stargazing from the center of a frictionless merry-go-round. If you keep your gaze on a single star to the east, it moves straight up, as it does for observers at the equator. Instead of seeing the stars twist, you see Earth twist around you. It is like hovering above the North Pole, except Earth does not spin as quickly. The closer to the equator you are, the less rapidly the world spins.

Foucault Pendulum

A pendulum in a clock swings back and forth within a single plane of motion. The mechanisms in the clock keep it forever tracing out the same path. However, if a pendulum is free to move in all directions, as a piñata can, the direction it swings will slowly rotate. Such a contraption, known as a *Foucault Pendulum*, swings back and forth without being materially affected by the friction holding the pendulum's base to Earth. Hence, to people watching the pendulum, the direction of the pendulum's path appears to change. In reality, the pendulum is not changing its direction: we are simply rotating around it.

The rope example I mentioned earlier works for the same reason. Here we use a pendulum because it gives us a way of seeing which way the pendulum is "facing" at a given time. We see the pendulum facing different places at different times because Earth is rotating underneath it.

A Foucault Pendulum. The pendulum swings in the same direction at all times; however, Earth rotates underneath it, so it swings along different marked lines throughout the day.

Moving mountains

Consider a person looking due south to Mount Hoverla, the highest mountain in Ukraine. She is looking straight ahead horizontally, neither inclining her head nor looking downward. Because she is north of the equator looking south, the point she sees straight ahead is moving more quickly eastward than she is. The circle made by that point as it moves around Earth is larger than the circle made by the woman north of it.

Even though the mountain is moving relative to her, the distance between it and the woman does not change. If an object is always moving to the left of an observer while staying the same distance from her, it can only be the case that the object is tracing out a circle (or part of one) relative to her. The mountain is moving around her, but she does not realize it because Earth is twisting her to always face the same direction.

Now consider the same person standing at the equator looking north at a mountain, say Mount Stanley in the Democratic Republic of Congo. Because she is exactly at the equator and looking due north, any point she sees is as far away from Earth's axis as she is (see facing page). This means the spot she is looking at is not moving relative to her.

A somewhat sloppier way of seeing this—which does not take into account geometry or the angle someone's face is tilted— is to think of two mountains equidistant from a woman in the Ukraine, one north, the other south. The mountain to the south is moving more quickly eastward, so its relative motion is to the left from her perspective. The mountain to the north is moving less quickly eastward, so its relative motion is also to her left. (Remember, she is facing the opposite direction when viewing it.) Thus, we can think of her as being in the middle of a circle with the mountains moving counterclockwise around her.

Consider the same setup centered on the equator. People on the equator are moving eastward faster than anyone else, so mountains to both the north and south are moving less quickly

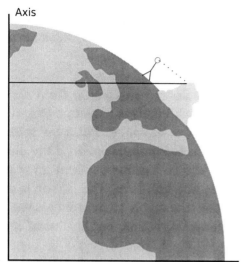

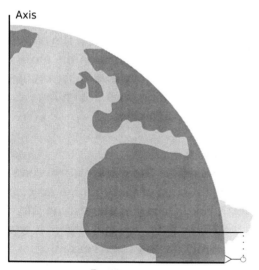

A person on the equator looking straight ahead can only see objects the same distance from Earth's axis, meaning she only sees objects rotating as fast as she is. A person anywhere else facing the equator sees objects moving faster eastward than she is.

eastward. The one to the north is moving to her left, but the one on the south is moving to her right. We cannot create a circle with the observer in the center and the mountains moving around her.

Seeing the world from space

A final way of seeing that Earth causes someone to spin is to see things from the perspective of space. Put a toy soldier on the equator of a globe and have him face north. If the globe has no axial tilt, this person will be looking straight up at the ceiling. Now, rotate the globe one-half turn. The person is still looking at the ceiling. His body (the arrow made from his feet to his head) is pointed in a different direction, but his line of sight is still directly upward.

Now do the same thing for someone at a middle latitude, say in Texas. The soldier will be looking in different directions after Earth rotates. If he first was looking up and toward one wall of the room, he will later be looking up and toward the opposite wall. The closer to the pole the soldier is placed, the more the direction he faces changes because his northward orientation is becoming less and less up/down (which does not change as he spins around) and more and more sideways.

Upshot

So, every person sees himself as the center of rotation on Earth (though the speed and direction of that rotation may vary), as though he is at the North Pole. Once you understand why the Coriolis effect works at the North Pole, you can accept it works everywhere (except the equator) for the same reason.

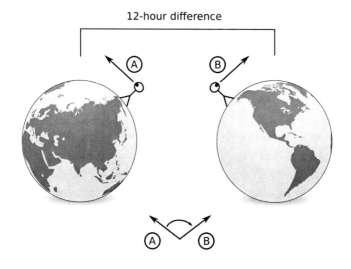

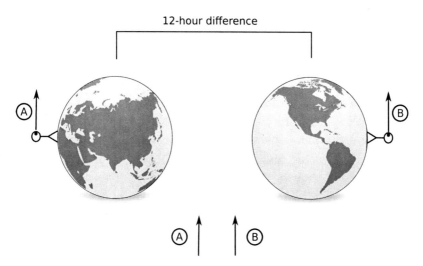

Someone in Alaska, at around 65 degrees latitude, is turned by Earth approximately 130 degrees in half a day. Someone on the equator is not rotated in this way; an observer facing the horizon looks in the same direction all the time. The amount Earth spins someone horizontally is equal to twice their latitude.

Interlude: The Full Story at the North Pole

We have found that Coriolis is easier to accept upon realizing that all non-equatorial points are similar to the North Pole. It pays, then, to develop our understanding of what is really going on with the hunter at the North Pole discussed on page 108.

The *Moving Target* analogy convincingly argues for Coriolis at the North Pole (where there is no east or west), but it does not consider the perspective of the arrow that is being deflected. The *Rotating at Different Rates* explanation is basically an effort to articulate the *Moving Target* explanation from the arrow's point of view. We will put that under a magnifying glass and see why saying "the rotation of Earth under the arrow changes during flight" masks an important nuance, a nuance that helps explain Coriolis for arrows shot in any direction.

Much can be learned by considering the difficulties that arise as we try to compensate for the difference in rotational motion of Earth under the arrow. We once again consider a hunter at the North Pole shooting a bear in Greenland, 1,200 miles away. The leftward motion of Earth is 0 mph at the hunter's location and increases to 300 mph upon arrival at the bear. At the halfway mark, the terrain is moving 150 mph to the east.

Let's say the hunter tries the simplest solution. He puts a rocket on the side of the arrow that fires on halfway through the trip. The rocket has just enough fuel to increase the arrow's eastward velocity by 300 miles per hour. This is twice the rotational velocity of Earth at that point. He hopes that adding twice the necessary velocity at the halfway point can make up for the arrow lagging behind for the first half.

The facing page shows the resulting flight.

It is probably not surprising that he missed the bear. He did, after all, try to deal with all the necessary curvature issues in one stroke. What is puzzling is that he did not even get close!

How might he do better? Let's say he tries to better match the changing speed of Earth's rotation by spreading out the

Time = 0 hours

Time = 3 hours

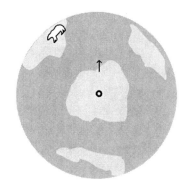

Time = 4 hours

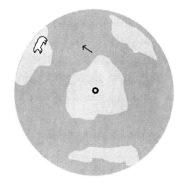

Time = 6 hours

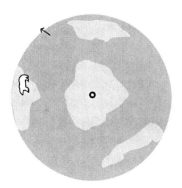

rocket power. Instead of having a single push of 300 miles per hour to the east, he uses 5 evenly spaced pushes of 60 miles per hour. This helps because each blast changes the velocity in the direction that the arrow perceives as eastward at the time rather than the direction the hunter originally viewed as leftward.

After this correction, the arrow still does not make it halfway to the bear. Even if he sets the rockets at just the right power so that its fuel is used up gradually over the entire trip, so the amount the rocket pushes the arrow to the east exactly matches the change in Earth's rotational speed, it would not bridge the gap. Though it is not at all obvious, it turns out such an ideal arrangement would get the arrow exactly halfway to the bear.

This frustrating result presents a paradox of sorts. It is not hard to see the basic problem. In the first example, the bear was moving to the left at 300 miles per hour the entire time, but the arrow was only moving to the left at 300 miles per hour for the last half. Even before correcting for the circular motion of the bear, that means the bear moved twice as far to the left as the arrow over the course of the journey.

This line of thinking makes us wonder whether the arrow could ever both match Earth's speed and still hit the bear. We are told the Coriolis effect is due to the arrow's traveling over regions rotating at different speeds, but even when we correct for this variation, hitting the bear seems hopeless. In oversimplified terms, the bear is always moving 300 miles per hour to the east, but an arrow moving at the same speed as the terrain under it will never be moving 300 miles per hour eastward until the very last instant. How can the arrow hit a bear that is always moving eastward faster than it is?

So we see this explanation masks some interesting subtleties. From one perspective, an arrow traveling with the same speed as Earth should appear (to the hunter) to be going straight. That view says the arrow hits the bear. On the other hand, the bear is always traveling eastward faster than such an arrow (except at the end) because the rotational speed at the bear's latitude is greater than at any point north of it. That observation suggests the arrow always misses to the west.

Paradox unraveled

The Coriolis effect essentially describes how spinning affects our judgment of motion. Since we are only treating horizontal motion, we care about two directions: left/right and forward/backward. The hunter who put rockets on his arrows was correcting for the effect on east/west motion. As the arrow moves forward, it must accelerate eastward to match Earth's rotation. What the rockets failed to address was the effect of rotation on forward/backward motion.

The direction we see as forward changes as Earth spins us.

The hunter at the pole is not going east, but he is spinning. Friction between his shoes and Earth's surface causes him to twist with Earth. Thus, the direction he perceives as frontward when he shoots is not the same direction he perceives as frontward after Earth has twisted him awhile. The arrow's initial velocity is in the direction the hunter used to be facing—not the direction he faces later.

For the arrow to continue in the direction the hunter sees as frontward, it must turn eastward constantly, above and beyond the modification for the different speeds of Earth's rotation. The two corrections turn out to be mathematically equal to one another, but they are both required.

To nail down the difference between these two separate corrections, let's say the hunter repents of his deadly sport, gives up on arrows, and buys a remote-control mini-helicopter he will use to send cookies to the bear as a peace offering.

We assume the trip still takes six hours and determine how much the helicopter's velocity has to change to stay on target. It begins going 200 miles per hour toward the bear, moving in the direction the hunter considers frontward.

Consider the question "How fast must the helicopter be traveling (at different times) as viewed from an astronaut above Earth?" We'll say the astronaut is facing Earth, looking straight down at the contrite hunter. We will further say that the astronaut's "up" direction is exactly equal to the direction the hunter was originally facing. In other words, the astronaut's perspective is nothing more or less than the perspective shown in the earlier figures with the bear.

According to the astronaut, the helicopter starts by moving straight "up" at 200 miles per hour. We will write this as $\langle 200,0 \rangle$. The first number says how fast upward the helicopter is flying. The second number says how fast to the left it is flying. Originally, it is not flying to the left or right, so that number is 0.

How fast is the helicopter moving when it makes it to the bear if the hunter managed to fly it straight there (from his perspective) with constant speed? Well, the hunter is facing toward the astronaut's left, and he still thinks the helicopter is moving at 200 miles per hour. This means that the left/right speed of the helicopter must be 200. Furthermore, the bear is moving downward due to Earth's rotation at a speed of 300 miles per hour. This means the speed in our notation would be $\langle -300,200 \rangle$. The first term is negative because the bear is moving in the opposite direction from the helicopter's initial motion. The astronaut sees the helicopter moving backwards and to the left compared with its original direction of flight.

And now we start to see the first hints of what went wrong with our efforts to fix the arrow with rockets. *The difference in rotational speeds does not equal the difference in velocity the arrow must attain.* The bear is rotating 300 miles per hour faster than the hunter (who is not rotating at all), but the change in speed required of the arrow is significantly more. The change in the up/down direction is 500 miles per hour. The change in the left/right direction is 200 miles per hour.

Some of that difference is due to the changing rotational speed at Earth's surface; the rest is due to the hunter's rotation; he constantly has to accelerate the helicopter to his left for it to continue in the direction he considers straight ahead.

It turns out that these two corrections are mathematically equal, but it is hard to see that when considering a 6-hour trip. Instead, let's look at just the first hour. We can write the velocity of the helicopter after one hour as:

$$V = I + S + E$$

Here V is the velocity after one hour, I is the initial velocity, S is the change in velocity to accommodate the spinning perspective of the hunter, and E is the eastward change to make sure the helicopter is moving at the same speed with everything else at that latitude. In other words, S takes care of the effect the hunter's rotation has on the helicopter's apparent forward velocity while E takes care of the effect the helicopter's motion has on its apparent side-to-side velocity. E is invisible to the hunter. Objects the hunter sees "at rest" in front of him are actually moving with a velocity of E.

Under normal circumstances (see boxed discussion below), E and S increase together because the helicopter is moving forward to new places at the same time that the hunter is spinning. They happen to be equal to one another, but it is easier to see this truth than explain it with words. With that in mind, I show

---

The difference between S and E

To see the difference between S and E, imagine God stopped time and nudged the helicopter forward along its route. Prior to starting time again, in order for the hunter to believe the helicopter was still moving straight, God would have to change the E component of its velocity (it is in a different location, so Earth is moving at a different rotational speed) but would not have to change the S component. Time was stopped, so the hunter did not spin any further. There is no need to compensate for the hunter's facing a different direction.

---

on the facing page the initial forward and leftward speeds in black and the speeds that must be achieved after one hour in gray, taking into account the fact that the hunter is spinning. We see that the sideways change in velocity required for the forward motion to stay forward relative to the hunter (S) matches the additional eastern velocity the helicopter will need to acquire so that the hunter sees it as going the same speed as everything else at that point (E).

This illustrates why an object must be pushed twice as hard to the left as one would expect based only on the difference in rotational speeds. Half of the acceleration is used keeping it facing the right direction, making up for the spinning observer, leaving only half to provide the extra rotation needed to keep up with Earth's surface as the arrow moves south.

Upshot

So, we see in the case of someone at the North Pole that there are two separate causes for the arrow's apparent deflection. The first is that different points seen as an observer looks south move eastward at different speeds. The second is that observers *spin with Earth's surface*, so the direction seen as "straight in front" changes over time.

It is not hard to see that the first cause will show up with any north/south motion. It is also clear that the second would manifest for someone standing at the Pole regardless of which direction the arrow was shot. A greater challenge lies in explaining why the first cause is not limited to just north/south motion and the second is not limited to the North Pole.

I have given one version of why these two effects happen everywhere: in some sense everywhere (other than the equator) is like the North Pole. My second explanation for why these two effects work everywhere in all directions is more concrete.

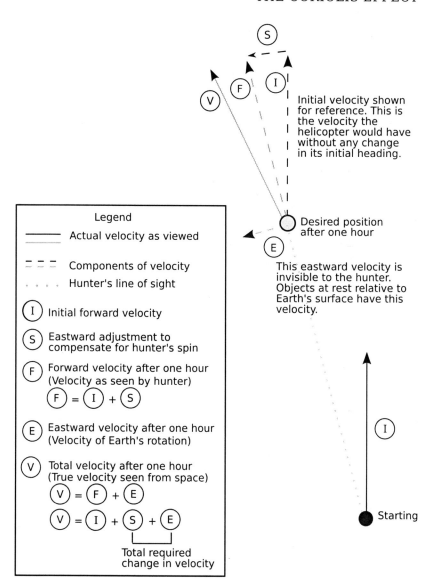

Initial velocity shown for reference. This is the velocity the helicopter would have without any change in its initial heading.

Desired position after one hour

This eastward velocity is invisible to the hunter. Objects at rest relative to Earth's surface have this velocity.

Legend

————— Actual velocity as viewed

– – – Components of velocity

· · · · Hunter's line of sight

(I) Initial forward velocity

(S) Eastward adjustment to compensate for hunter's spin

(F) Forward velocity after one hour (Velocity as seen by hunter)
$$(F) = (I) + (S)$$

(E) Eastward velocity after one hour (Velocity of Earth's rotation)

(V) Total velocity after one hour (True velocity seen from space)
$$(V) = (F) + (E)$$
$$(V) = (I) + (S) + (E)$$
Total required change in velocity

I

Starting

The change in velocity the object must acquire due to its new location (E) matches the change in velocity required to account for the hunter's spinning (S). In this picture they have slightly different directions because we have approximated the instantaneous push required by looking at two positions an hour apart.

Account 2: Curving Latitudes

One can explain why the Coriolis effect works for all objects moving in any direction, and in particular objects moving due east or west, by a close examination of latitude lines.

Imagine looking at Earth from space. Each point on Earth, as it rotates about the axis throughout a day, makes a circle centered on Earth's axis. The North Pole's circle is just a dot. Someone standing in Ontario makes a smaller circle than someone standing in Texas. The person standing in Texas traces out a circle smaller than someone standing on the equator in Ecuador. Since all of these circles are traced out in the same time interval (24 hours), the person in Ontario is moving more slowly eastward than the Texan, who is moving more slowly eastward than the Ecuadorian. The basic rule is that the farther one is away from Earth's axis, the bigger that person's circle is and the faster she is moving toward the east.

One could easily get the impression that if you shot an arrow at a bear who was on the same circle (which is really just a given line of latitude), there would be no Coriolis effect because the bear would be traveling the same speed (eastward) as you are. There are two problems with this conclusion, and both of them spawn from the observation that *there is no such thing as "going east in a straight line" unless you are on the equator.*

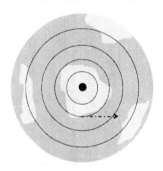

Moving in a straight line near the North Pole quickly takes you off your original line of latitude.

To see this, let's take an extreme case. Imagine you are very near the North Pole. It is easy to see that, even if you start out going due east, you will soon find yourself traveling away from the North Pole if you continue forward without turning. In order to stay on the same latitude, you constantly have to turn to the left as you move forward.

You can see this easily if you have a small toy car and a ball. Put the car anywhere on the ball with any initial direction you want, and roll it without turning. The trip it takes will always be a *great circle*, a circle that cuts the sphere into two equal parts. The equator is a great circle, and so is any other circle having the same diameter.

If you do not have a toy car handy, you can do a thought experiment. Pretend you are on a motorcycle with a wide, rolling paintbrush for a back tire. If you drive your motorcycle straight, you will eventually go all the way around the world, ending up back where you began. Now, cast a magic spell transforming the paint left by your tire into a satin ribbon.

Objects moving without steering move in great circles. Half of a great circle is shown above. The other half is on the obverse side.

Since you didn't turn to the left or right, the left edge of the ribbon must have the same length as the right edge of your ribbon. If you had turned, the ribbon's inner edge would be slightly shorter than its outer edge. The only way to put such a ribbon snug around Earth is for the ribbon to be a great circle.

Thus, if you want to move due east and you also want to move in a straight path, the only option you have is to move along the equator. All other eastward paths require turning. This observation holds the key to the greater mystery of why the Coriolis effect is the same strength in all directions.

The situation is well-depicted on gnomonic projection maps, which portray the curved lines of latitude. Imagine you are at

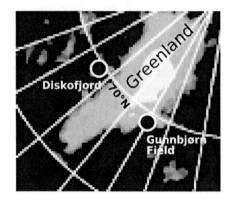

70°N latitude, just north of Diskofjord, Greenland. You are shooting at a bear, also at 70°N latitude near Gunnbjørn Fjeld. There are three motions to consider. First is the bear's motion along the 70°N line (due to rotation of Earth). The second is your own motion along the same latitude. Third is the direction the arrow is traveling, which does not change. Once the arrow is released, it appears to curve for two separate reasons:

- What is "east" to you is not parallel to what is "east" to the bear. This is similar to the first example where a hunter shot south from the North Pole. In that example, the bear was moving at a different speed. In this example the bear is moving in a different direction. The effect is the same.

- You continue to move along your line of latitude as the arrow flies. Since this line of latitude is not straight, what you consider "straight ahead" is no longer the same direction you considered "straight ahead" when you loosed the arrow. This is equivalent to the twisting effect more easily seen by someone at the North Pole being spun about the axis under foot.

These two effects combine to make the arrow deflect even though you were shooting at something on the same latitude, but it is not hard to see that they apply for any direction. If you were shooting northeast or some other direction, we would replace the first point above with a combination of two factors, one being from the north/south component (due to difference in relative speeds), and the other being from the east/west component (due to moving in different directions). The second bullet, of course, applies without modification since you are moving along your line of latitude regardless of what direction your arrow is flying.

Conclusion

Earth-bound observers see moving objects curve for two separate and equally important reasons. "Straight paths" on Earth's surface (i.e. paths made by objects that are not curving right or left) form great circles. This means that an object moving straight will always pass over terrain rotating at varying speeds unless the object is going due east or due west around the equator.

The second reason objects appear to deflect is that an observer at any point other than the equator is being slowly turned around by Earth's rotation, so a freely flying object that appears headed in one direction at one moment will appear headed in a different direction the next.

Addendum: Coriolis in Real Life

The Coriolis effect is often called a "fictitious force," much like centrifugal force, because the deflection is illusory. To someone in space, a thrown object will take a straight path. It only appears to an observer on Earth to curve because the observer is being spun about by Earth's rotation. This is similar to how someone watching kids on a spinning merry-go-round will see a ball rolled across it move straight, while the children on the merry-go-round will see it curve. Yet there is nothing illusory about hurricanes, ocean currents, or other phenomena influenced by Coriolis. A natural question to ask is how a fake force like Coriolis can have real ramifications.

The short answer is that the Coriolis effect has real consequences *whenever it is coupled to a real force that rotates with
the Earth*. The Coriolis effect modifies a moving object's trajec-
tory relative to Earth's preferred coordinate system (latitude
and longitude), so it can have a real effect on an object being
acted upon by forces linked to Earth's surface.

For example, the Sun's heating causes air near the equator
to rise, creating a low-pressure region. This natural distribution
of pressure produces a force based on latitude. Because the
force felt by a particle depends on its position relative to Earth's
surface—in this case its latitude—we have ripe conditions for
Coriolis to have a real effect.

After rising, the warm, moist air moves away from the equator, toward the poles. This pulls cooler air in the lower atmosphere toward the equator. The air moving away from the
equator curves due to Coriolis. By the time the particles reach
30 degrees latitude, they have turned all the way around and
begin moving back to the equator.[6] This sets up three basic wind
patterns, or *cells*, in each hemisphere: one near the equator, one
near the pole, and one in between. Without the Coriolis effect,
there would just be one cell in each hemisphere going straight
from equator to pole and back.

Cyclones pose a more confusing example. Students are told
repeatedly that hurricanes and typhoons twist counterclockwise
in the Northern Hemisphere and their counterparts turn clockwise in the Southern Hemisphere, yet the Coriolis effect would
appear to cause the opposite. If flying objects, like moving air
particles, curve to the right in the Northern Hemisphere, the
natural expectation is that that hurricanes there twist clockwise.

A cyclonic storm (e.g., a hurricane) is caused by a low-
pressure region. Low-lying air moves with Earth's surface, so

---

[6]They lose altitude as well so the cooler air coming back to the equator
does not fight against the warm air flowing away from it.

this low-pressure region is tied to Earth's coordinate system. Air is naturally pushed from high pressure to low, so we have another example of a genuine force rotating with Earth's surface.

Let's consider what would happen if the Earth did not rotate. When there is high air pressure in one region and low pressure in another, air simply rushes from the first to the second, balancing the pressures.

But the Earth does rotate, *and the low pressure region rotates with it.* This means the air particles will miss their target, deflected by the Coriolis effect. You can think of the hurricane's eye as an observer that sees all the particles rushing toward it curve sideways. The air is coming into the cyclone from outside so it deflects to the right (in the Northern Hemisphere); this constant deflection causes a counterclockwise circulation around the center.[7]

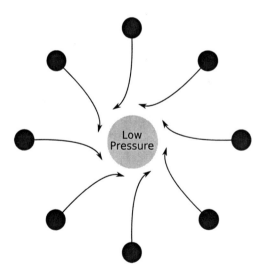

<hr />

[7] Incidentally, while hurricane winds spiral counterclockwise due to rightward deflection of air coming from the outside, the *clouds* that are spewed out from the top of a hurricane spin clockwise—the key difference being that they are moving away from the eye rather than toward it.

Therefore, Coriolis causes the air particles to curve away on their route to the low-pressure region. They spiral around and around the center rather than entering it, which would raise the pressure in the center, neutralizing the hurricane. Since hurricanes depend on Coriolis in this way, it is *extremely* rare for them to form near the equator, where the Coriolis effect is very weak.

For readers who wish to learn more about any of these topics, I'm providing notes to each one. I include vocabulary terms to help any interested parties do their own further inquiry.

Chapter 1: The Seat Buckle Sting: Why Do Metals Feel Hot?

The scientific term for sensing heat or cold is *thermoception*.[8] The discussion in the chapter focused on how short-term changes in temperature affect our body. The long-term effects of heat or cold constitute a more complicated topic involving, among other things, the transfer of heat by our bloodstream. Anything affecting fluids in the body, particularly blood or air,[9] can significantly alter the body's temperature regulation (*thermoregulation*). For example, drinking alcohol heats us up by increasing blood flow (*vasodilation*); divers plunging into some environments require pre-heated air.

The most basic idea in thermodynamics is "heat moves from hotter objects to colder ones." This is typically known as the *Zeroth law of thermodynamics*, even more fundamental than the *First law of thermodynamics*, which is "energy can be neither created nor destroyed." The scientific measure of how well a material offloads or receives heat is the *heat transfer coefficient*. For the specific case of two different solid materials touching each other, the term is *thermal contact conductance*.

Metals conduct heat for the same reason they conduct electric current: their electrons are free to move about, transporting

---

[8]Scientific terms for other senses include *nociception* (sense of pain), *proprioception* (sense of bodily orientation, in particular the configuration of our limbs in space), and *equilibrioception* (sense of balance).

[9]A reminder: "fluids" in scientific parlance refers to both gases and liquids.

energy from one location to another. However, this connection does not extend to non-metals: diamond has the highest thermal conductivity of any naturally occurring material, but it does not conduct electricity well. Jewelers can determine if a mineral is a diamond or a fake by measuring how quickly it conducts heat.

Chapter 2: Cloud Formation

Clouds in a Glass of Beer: Simple Experiments in Atmospheric Physics by Craig F. Bohren has a lot of information on cloud formation and other fun stuff. Alistair B. Fraser, Emeritus Professor of Meteorology at Pennsylvania State University, has a great webpage on various commonly mistaught items, including cloud formation. His material is available at http://www.fraser.cc. Steven Babin, Senior Meteorologist, Physician, and Engineer at Johns Hopkins University Applied Physics Laboratory wrote a nice article on the misrepresentation of relative humidity, using much of Bohren's material. That paper, *Water Vapor Myths: A Brief Tutorial*, is available at http://fermi.jhuapl.edu/people/babin/vapor/index.html. I am indebted to all three of these scientists for my own presentation.

If you put water in a jar and then close it up, an *equilibrium* will quickly arise between the water and the water vapor in the air above it. Water will constantly be evaporating out of the liquid into the gas, and water vapor will constantly be condensing out of the gas into the liquid. The immigration and emigration balance, so the total amount of water in the air is stable.

If you raise the temperature of the water, the rate of evaporation will increase, meaning that the air becomes more humid. However, since there is more water vapor in the same volume, condensation will also increase. At some point, the air in the jar will get humid enough that the elevated rate of condensation will match the elevated rate of evaporation. In other words, equilibrium will be re-established.

136

The pressure of the water vapor in such a situation—a closed container where both the vapor and liquid exist in equilibrium—is known as the *vapor pressure*[10] for that temperature. Most confusion about relative humidity can be attributed to not understanding this definition. Relative humidity compares the pressure of water vapor in the outside air with the vapor pressure for that temperature. That is to say, relative humidity is always a comparison of air's current water content to the maximum possible in a closed container at the same temperature *that already has liquid water in it.*

It is much easier for water vapor to condense onto a watery surface (like a lake) than it is for it to condense onto a dry surface, and it is much, much easier for it to condense onto any surface than it is for it to condense in mid-air. Hence, in no regard can 100% relative humidity be accurately thought of as the maximum amount of water vapor that "the air can hold," even if we set aside the other problems revolving around that phrase.

Chapters 3–4: Genetics and Evolution

The writings of naturalists around the turn of the 20th century provide a great starting place to understand the interplay between simple models of genetics and evolution. I highlighted Thomas Hunt Morgan's work, and his *A Critique of the Theory of Evolution* is an accessible and informative read. This work and others give a fantastic window into the thoughts of the world's

---

[10]Because this pressure is the amount occurring at equilibrium, it is sometimes called *equilibrium vapor pressure.* Furthermore, because any greater pressure would lead to higher condensation and an eventual return to equilibrium (assuming a closed vessel with a surface of liquid to condense into), it can also be called the *saturation vapor pressure*, but that is a poor choice of words because it leads people to think the **air** is saturated when the (dry) air has almost nothing to do with evaporation. (The pressure of the air does affect boiling temperature, but that is a separate topic.)

sharpest scientists working within a model similar to the one currently taught to students.

To move forward from that point, I recommend Stephen Jay Gould's magisterial *The Structure of Evolutionary Theory*. The first part of that tome presents nearly 500 pages discussing the history of evolutionary thought. The second part discusses critically the present state of evolutionary theory. Of course, this is all from Gould's point of view, but he was a giant—more than a giant—in the field.

A variety of resources defend different views on the theory itself. Richard Dawkins is perhaps the world's most famous militant neo-Darwinist (and atheist,[11] for that matter). His books, such as *The Selfish Gene*, provide suggested modifications to traditional neo-Darwinism[12] to account for its classical problems. A different kind of patch, offered by another devotee to traditional neo-Darwinism, is presented by David Sloan Wilson in his books, which stress group dynamics.

Further afield from these views are Gould's own. In addition to *The Structure of Evolutionary Theory* cited above, three books (*Ever Since Darwin, The Panda's Thumb, Hen's Teeth and Horse's Toes*) offer compilations of articles written by Gould for *Natural History* magazine. Kim Sterelny's *Dawkins vs. Gould: Survival of the Fittest* is an informative reference describing Gould's heterodoxy. It is well written but consistently understates the philosophical differences between the two camps.

---

[11] Dawkins uses this term, "militant atheism," to describe what is required by all scientifically minded people today to push back the scourge of religion and its effects on science education. A video capturing a long presentation made at the invitation of TED.com discussing the topic can be found at http://www.ted.com/talks/lang/eng/richard_dawkins_on_militant_atheism.html.

[12] I use the phrase "traditional neo-Darwinism" to refer to the version that has evolved today, which is dominated by natural selection, as opposed to the original, more multi-faceted approach conceived in the 1930s. Thus "traditional" neo-Darwinism is, ironically, not really traditional at all.

Lynn Margulis has written several books describing her heretical, symbiogenesis-centered view. She has also championed the *Gaia hypothesis*, which views the entire Earth as an organism. Both her ecological and evolutionary claims are presented in *Symbiotic Planet: A New Look at Evolution*.

Lee Spetner comes from a different angle in *Not by Chance: Shattering the Modern Theory of Evolution*. It is an interesting, informative, and involved discussion of the problems behind traditional evolutionary theory from the standpoint of information theory. Most of the book constructs an argument for why traditional neo-Darwinism cannot explain the genetic complexity found in modern DNA. His argument requires a long attention span to appreciate, but is rewarding once completed. En route, the reader is provided a large cache of information about the biochemistry of genes and the importance of separating structural genes from regulatory genes.

Spetner presents *directed mutation* as a solution to the various problems. This controversial view claims that cells, through non-naturalistic means, can somehow sense the mutations required to adapt to a new environment. In this, he is proposing a more scientifically up-to-date version of an evolutionary theory championed by many in the late 19th century. Those who advocate this model have encountered a great deal of resistance, and skeptical scientists have published papers aimed at refuting the experimental evidence supporting it.

The case of nylon-eating bacteria is interesting because it gives a verified laboratory example of a mutation causing natural selection by altering a *structural* gene. Some have questioned its importance because it makes the enzyme in question less specific rather than more specific. Skeptics claim this "degeneration" (as they call it) of the enzyme generally makes an organism less fit in natural settings and does not explain where the original ability came from. In this case, bacteria already had an enzyme that digested certain kinds of chemicals, and the mutation tweaked the exterior of the enzyme so that nylon could latch onto it. Though it bears a cosmetic resemblance to Spetner's argument, I do not find this perspective very compelling, at least based on the current understanding of this specific mutation. Interested readers should not have problems finding articles written from both perspectives on this controversial example.

Evolution has become such a multi-disciplinary field that it is difficult for any faction to appreciate or even understand others' points of view. The traditional version emphasizes, understandably, the fields that existed when neo-Darwinism spawned. Scientists in more recently developed subfields have had trouble gaining a fair reception from those whose allegiance is with the traditional view. These disenfranchised scientists have been the most interested in changing the model, and researchers in a wide variety of fields convened in Altenberg, Austria during the summer of 2008. *Nature* ran an article, "Biological theory: Postmodern Evolution?," in the subsequent September issue. Suzan Mazur wrote a book about the meeting, *The Altenberg 16: An Exposé of the Evolution Industry*, but I cannot recommend it with any enthusiasm. Beyond the understandable scientific agenda, the author appears to have a deep-seated political one, attacking the implicit justification of war that she and others find in the acceptance of Darwinism.

She is certainly not the first to have suggested this link. That Hitler used Darwinist rhetoric to justify war is well known; in chapter XI of *Mein Kampf*, he speaks repeatedly and at length about how it is only natural for the stronger races to destroy the weaker ones,[13] claiming the "process of development towards a higher quality of being" would cease if inferior specimens

---

[13] "The stronger must dominate and not blend with the weaker, thus sacrificing his own greatness. Only the born weakling can view this as cruel, but he after all is only a weak and limited man; for if this law did not prevail, any conceivable higher development of organic living beings would be unthinkable."

"possessed the same capacities for survival and procreation of their kind." Sadly, Hitler was not alone in this perversion of Darwinism, believing that humanity should support or even accelerate the "survival of the fittest" within its own ranks. Many intellectuals of the era, including several evolutionary biologists, agreed with Hitler at least in principle if not in means.[14] Indeed, it was the connection between Nazism and eugenics that made the latter unfashionable after WWII.

Fortunately, this whole topic is merely of historical interest.[15] Logically speaking, it is as intellectually bankrupt to attack Darwinism by linking it to war, Nazism or the eugenics movement as it is to attack faith by linking it to various atrocities humans visit upon one another in the name of God. Thus, Mazur manages to perpetrate in her attacks on Darwinism the same error Dawkins perpetuates in his defense of Darwinism against the negative outworkings of theism.

More notes on the history of evolution and non-gene-based hypotheses (e.g., intelligent design and creationism) can be found in the *Notes* chapter of Volume 3.

Chapter 5: Why Are Veins Blue?

The main source material for this chapter was *Why Do Veins Appear Blue? A New Look at an Old Question*, an article by Kienle et al. in *Applied Optics* (Vol. 35, Issue 7, pp. 1151ff). It is a rather technical read. If you are interested in getting into the nitty-gritty of it, you should first read up on absorption

---

[14]It was the support by scientists in the U.S. that compelled lawmakers to pass forced sterilization programs on the west side of the Atlantic. *In the Name of Eugenics* by Daniel J. Kevles has three times more than anyone should possibly want to know on this unsettling topic.

[15]Another book on this controversial topic is *From Darwin to Hitler: Evolutionary Ethics, Eugenics, and Racism in Germany*. It is very heavily suggested, though, that anyone choosing to read this book first read Darwin's own writings to better judge the tenability of the author's accusations.

spectra. Key terms here are *molar extinction coefficient* and *absorption coefficient*. These describe how easily a material absorbs photons of different frequencies.

A related topic that requires less technical knowledge is the study of light filters. There are plenty of interesting gems to be found in that mine. For example, you can trick your eyes into seeing infrared light by filtering out as much of the "visible" spectrum as possible. When the colors we normally see are blocked, our eyes become sensitive to frequencies that otherwise get drowned out. With these self-made night vision goggles, you can see through the outer layers of many plants, and the whole world takes on a different look.[16]

In the chapter I point out that veins appear blue against fair skin owing to an optical illusion of sorts. Our brain is tricked into seeing blue owing to the contrast of the vein with the neighboring skin. This phenomenological curiosity is known as *color constancy*.

Scattering is the general term for light bouncing around as it encounters molecules. *Rayleigh scattering* is the specific type giving the sky its blue hue. Both Rayleigh scattering and a more complicated type (*Mie scattering*) are relevant for light scattered by skin. The *optical window* refers to the frequencies of light that can travel the farthest in organic tissue before being absorbed. Since fatty tissue has different absorption properties than lean tissue, some have suggested using radiation testing as an alternative method of grading the leanness of meat.

Chapter 6: Producers and Consumers

Ecologists have a variety of classifications for organisms. The producer/consumer distinction refers to where an organism

---

[16] Any conscientious presentation of such homemade goggles will warn the user not to look directly at the Sun, or will provide directions protecting against UV light.

gets its carbon. To distinguish organisms that get their energy from light from those that get energy from matter, the terms *phototroph* and *chemotroph* are used. To classify organisms by what they use for fuel, *organotroph* and *lithotroph* are used. Organotrophs use organic compounds for fuel; lithotrophs (literally "rock eater") use inorganic compounds. It is important to note that plants are lithotrophs because water is their primary fuel.

I am not using "fuel" loosely. All organisms need a source of electrons for their metabolism. This source of electrons (called a *reducing agent*) is what I refer to as fuel. Green plants split water molecules apart using energy from the sun in a process called *photolysis*. The electrons from that reaction are used later to drive photosynthesis. Since water is an inorganic compound, green plants are categorized as lithotrophs, as are the bacteria that feed on ammonia, hydrogen sulfide, or hydrogen gas.

These latter specimens, known as *chemosynthetic bacteria*, are a fascinating and diverse bunch. Many are found near *hydrothermal vents* on the ocean floor. Organisms that live there and in other atypical environments are known as *extremophiles*. As there are a wide variety of such "extreme" climates, the organisms in this class are diverse.

Chapter 7: The Far Tide

A much less technical description of tides can be found in Phil Plait's *Bad Astronomy* book. His chapter encompasses more facts and trivia about tides but does not go as deeply into the guts of the far tide and how geometry and gravity conspire to cause tides on Earth. For those interested in the other side of the technical spectrum, read Mikolaj Sawicki's[17]

---

[17]Sawicki has a webpage (http://www.jal.cc.il.us/\~mikolajsawicki/bad_physics.htm) dedicated to published errors in science and mathematics.

"Myths about Gravity and Tides" (*The Physics Teacher,* October 1999). A revised version is available at http://www.jal.cc.il. us/~mikolajsawicki/tides_new2.pdf.

Donald Simanek has an accessible presentation of tides at http://www.lhup.edu/~dsimanek/scenario/tides.htm. He does a good job of illustrating the common myths and misconceptions about tides before presenting a gravity-based answer.

At a certain latitude, the tidal force changes from pushing outward to pushing inward. This is depicted in the figure on page 83, which is inspired by Barger and Olsson's *Classical Mechanics, a Modern Perspective.* Determining what that critical latitude is makes a good problem for a bright AP physics student. The problem is challenging on many levels, not the least of which being that it can be approached in many ways, some far easier than others. For anyone who cares, the answer is approximately $54.7°$[18].

Chapter 8: The Greenhouse Effect

Professor Fraser, whose meteorology page I referenced in the notes to the *Cloud Formation* chapter, also has a page devoted to the greenhouse effect. In particular, he gives a good explanation of why our atmosphere is best seen as a separate heat source for Earth rather than conflating the atmosphere and planet into a larger system. That page is located at http://www.ems.psu.edu/~fraser/Bad/BadGreenhouse.html.

An excellent source for information about the Moon and how it compares with Earth is *Lunar Sourcebook: a User's Guide to the Moon* by Heiken, Vaniman, and French. If you want to learn more about the specific physics behind why planets with moderated temperatures are less efficient radiators than those

---

[18]This is the angle required to make a right triangle with side lengths of 1, square root of 2, and square root of 3.

with extreme temperature ranges, you will need to study the *Stefan-Boltzmann Law.*

Studying an "energy budget" is a useful way to see all the ways heat moves to and from Earth. Vaclav Smil presented a detailed energy budget in *Energies: An Illustrated Guide to the Biosphere and Civilization* (MIT Press, 2000). NASA has it posted at http://www.nasa.gov/images/content/57911main\ _Earth_Energy_Budget.jpg, and I have reproduced it below:

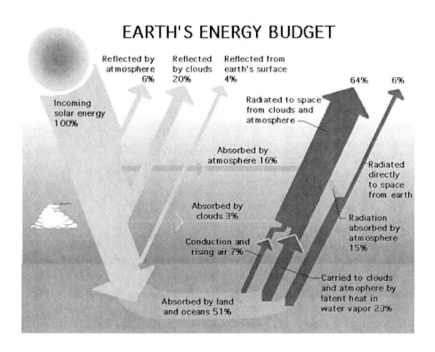

I like this budget because it separates the components well. It implicitly presents a mild "greenhouse effect" in the guise of the 15% on the right that gets funneled back into the atmosphere. Incidentally, this graphic shows up in a nice article describing how little we really understand about the effect of clouds on the atmosphere. That article is at http://science.nasa.gov/science-news/science-at-nasa/2002/22apr_ceres.

A famous portrayal of Earth's energy budget that makes the greenhouse effect more obvious is Trenberth, Fasullo, and Kiehl's version, recently updated in 2008 from a paper at http://www.cgd.ucar.edu/cas/Trenberth/trenberth.papers/BAMSmarTrenberth.pdf.

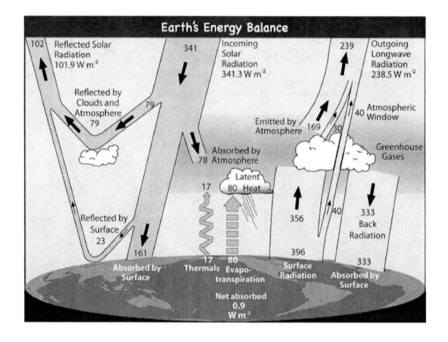

This figure shows the "back radiation" explicitly. I find it interesting that these two figures do not appear to agree. It is hard to imagine that "back radiation" accounts for over two thirds of the total radiation absorbed by the surface, as the second suggests, if the atmosphere receives less energy from the Earth (15%) than it does directly from the Sun (16%) as indicated in the budget posted on NASA's site.

It is not hard to find webpages dedicated to one or another view on global warming, more formally called AGW (Anthropogenic Global Warming). I can offer a little help in at least pointing out some of the higher-quality sites on each side of the debate.

A quality site dedicated to the pro-AGW side is `http://www.realclimate.org`. A less technical site is `http://www.skepticalscience.com`. Don't let the name fool you; their slogan is "Getting skeptical about global warming skepticism."

Sites that are definitely unconvinced about AGW include *The Air Vent*, located at `http://noconsensus.wordpress.com` and Steve McIntyre's `http://www.climateaudit.org`. The latter tends to have rather technical articles. McIntyre is an interesting character in the AGW debate. He has spent many years on a single mission: to get the raw climate data and models out in the open for debate and make sure that those making claims about global warming are not tinkering with the data.[19]

In addition to these four sites, which all have many links to other sites, you might be interested in the neutral ground at `http://www.climatedebatedaily.com`, a site aimed at more constructive dialogue.

Chapter 9: The Coriolis Effect

Mathematically, the size of the Coriolis effect equals twice the *cross product* of the velocity vector of the moving objects and the *angular velocity vector* describing how fast you are spinning around.

In layman's terms, this means the effect is greatest for fast-moving objects or for objects that are moving perpendicular to

---

[19] Intra-departmental discussion about his efforts to get original data from scientists show up in many of the emails leaked/hacked in the Climate Research Unit email controversy. The relevant personnel were told to ignore any Freedom of Information Act request originating from McIntyre.

the axis of revolution. The effect is zero if something is moving parallel to the axis. For example, an object moving due north at the equator is moving parallel to Earth's axis, so there is no Coriolis effect. Similarly, an object moving straight up from the Earth at the North Pole is also moving parallel to Earth's axis, so there is no Coriolis.

This makes sense. If you think of a rocket shooting straight up from the North Pole, it is easy to accept that an observer at the North Pole would see it shoot straight up without any curving. The same is true for all other observers moving with Earth's surface.

Someone standing on the equator, though, and shooting a rocket straight up would see the rocket curve quickly to the right. Indeed, a rocket shooting straight up from the ground at the equator is similar to a rocket shot horizontally from the North Pole. They are both perpendicular to Earth's axis.

About the Author

David Rudel has served since 2005 as senior editor for Explore-Learning, an acclaimed provider of science and math education software based in Charlottesville, VA. During that time Explore-Learning has won the CODIE award for Best Science Instruction Solution three times and has won the Association of Education Publisher's Award of Excellence three times along with AEP's Golden Lamp award. The last of these awards specifically honored their curriculum materials, which Rudel helps write. ExploreLearning has also won a Webbie for Best Education Website and won the CODIE for Best K-12 Instructional Solution (all subjects) in 2009.

Outside his role at ExploreLearning, Rudel is a published mathematician, professional math modeler, chess writer, and theologian. In the past he has taught middle school, lectured at Dartmouth, and (during his younger days) successfully competed in several science and math contests. While at Harvey Mudd College, he led a team garnering a world championship in COMAP's *Mathematical Competition in Modeling* and was also on their three-member Putnam Competition team, which achieved the highest rank among undergraduate institutions in 1997. Prior to transferring there, he twice joined fellow students at Grinnell College to win the Iowa Collegiate Math Competition. Competing individually, Rudel also won Texas state championships in both mathematics and physics.

Breinigsville, PA USA
12 December 2010
251189BV00005B/1/P